MUTUALISM: ANTS AND THEIR INSECT PARTNERS

A mutualism is an interaction between individuals of two different species of organism in which both benefit from the association. With a focus on mutualisms between ants and aphids, coccids, membracids and lycaenids, this volume provides a detailed account of the many different facets of mutualisms. Mutualistic interactions not only affect the two partners, but can also have consequences at higher levels of organization. By linking theory to case studies, the authors present an integrated account of processes and patterns of mutualistic interactions at different levels of organization, from individuals to communities to ecosystems. Interactions between ants and their insect partners and their outcomes are explained from a resource-based, cost–benefit perspective. Covering a fascinating and growing subject in modern ecology, this book will be of interest to community and evolutionary ecologists and entomologists, at both research and graduate student level.

BERNHARD STADLER is a Research Associate in the Department of Animal Ecology at the University of Bayreuth, Germany.

TONY DIXON is Emeritus Professor in the School of Biological Sciences at the University of East Anglia, UK, and author of two previous books with Cambridge University Press: *Insect Predator–Prey Dynamics* (2000), and *Insect Herbivore–Host Dynamics* (2005).

MUTUALISM

Ants and their Insect Partners

BERNHARD STADLER

Bayreuth Institute for Terrestrial Ecosystem Research,
University of Bayreuth, Germany

ANTHONY F. G. DIXON

School of Biological Sciences, University of East Anglia,
Norwich, UK

CAMBRIDGE
UNIVERSITY PRESS

CAMBRIDGE UNIVERSITY PRESS

Cambridge, New York, Melbourne, Madrid, Cape Town, Singapore, São Paulo, Delhi

Cambridge University Press
The Edinburgh Building, Cambridge CB2 8RU, UK

Published in the United States of America by Cambridge University Press, New York

www.cambridge.org
Information on this title: www.cambridge.org/9780521860352

First published 2008

Printed in the United Kingdom at the University Press, Cambridge

A catalogue record for this publication is available from the British Library

ISBN 978-0-521-86035-2 hardback

Contents

Preface

Mutualism, a reciprocal beneficial interaction between organisms, involves ecological and evolutionary processes occurring at several scales of organization. For many decades mutualism was the stepchild of ecology, neglected, malnourished and not studied theoretically because the prevailing paradigm was 'nature red in tooth and claw'. Positive interactions appeared to be both more difficult to accept on theoretical grounds and to test experimentally than negative interactions. In particular, trying to understand the conditions for stability and long-term outcome of positive interactions was (and still is) a great challenge. Now it is appreciated that there exists a continuum in the interactions between individuals belonging to different species, like ants and their insect partners, and this raises several interesting questions. For example:

- How can the different life histories of ants and their partners be merged so that interactions become beneficial?
- How does the abundance of the partners affect the strength of these interactions?
- What are the population and community consequences of mutualistic relationships for the interacting partners and indirectly affected species?
- Do mutualistic interactions affect species diversity?
- How does the environment affect the outcome and stability of these associations?
- How can the different partners of ants coexist in local and regional communities?
- How can mutualists persist in the face of exploiters?

In this book we aim to explain the underlying mechanisms of the dynamics of these associations by adopting a view that is not ant-centred, because the selection pressures of such associations are likely to affect both partners. As a consequence, we are more interested in describing the outcome of associations, which could be either negative or positive, or negative and positive at different times. In particular, we start with an historical perspective of mutualism and the theory of two-species mutualistic interactions. The differences

vii

and similarities in ecological traits of membracids, lycaenids and coccids when associated with ants and the many associations that have evolved between aphids and ants are addressed. Mutualistic interactions not only affect the two partners but can have consequences for higher levels of organization, such as communities and ecosystem processes, which are addressed in later chapters. Finally, the environmental and ecosystem problems that arise from mutualistic interactions, especially in combination with alien species, are addressed.

With regard to the breadth of the subject, the book should be of interest to community and evolutionary ecologists, and especially entomologists studying ants and their relationships with lepidopteran or homopteran partners. In addition, it provides students at the graduate level with a theoretical and experimental background to mutualism, which is a fascinating and growing aspect of ecology.

We gratefully acknowledge the support of the following institutions, which provided us with the opportunity to work on this complex subject by offering their resources in the form of access to literature, funds for travelling, housing and office space. In particular, BS would like to thank the Harvard University, Harvard Forest, for the generous support during a Bullard Fellowship. Having access to their unmatched literature resources was essential in the early phase of the project. The University of Bayreuth was his scientific base for many years while a student and scientist, and allowed him the freedom to follow his interests. On a more personal level, Pavel Kindlmann, with unprecedented generosity, provided a refuge and office space at a critical period in BS's career. Without his help the book would not have been completed. AFGD is particularly indebted to Professor Helmut Zwölfer who initiated the studies resulting in this book when he invited him to Bayreuth to meet BS at the time of writing his Ph.D. thesis, and to the European Science Foundation for awarding BS the Fellowship that enabled him to visit the University of East Anglia and initiate the studies into the costs for aphids associated with mutualism. We are also indebted to BITÖK/ University of Bayreuth for funding several visits of AFGD to Bayreuth, which greatly facilitated the exchange of ideas that resulted in this book.

1

The scope of the problem

The concept of natural selection proposed by Charles Darwin and Alfred Russel Wallace rests on the assumption that environmental conditions determine how well particular traits of organisms are suited for reproduction and survival. In this respect it is a conditional theory, which suggests different outcomes in different situations. That is, as long as the conditions remain the same, particular traits might continue to be adaptive and eventually become more common in a population. Changes in ecological conditions, which might be either bottom-up or top-down from the perspective of phytophagous insects, can drastically change the requirements and make previously well suited traits maladaptive. As a consequence, classifying interactions between different species as competition, predation, parasitism, mutualism, and so on risks being an oversimplification because of the ongoing changes in ecological conditions, which might continuously shift the nature and outcome of interspecific interactions. Bronstein (1994b) criticized the static view because it obscured the ecological and evolutionary links between the different interactions. In a dynamic world there are no fixed categories. However, placing interactions into different categories does help us understand at least pair-wise interactions, which, historically, have focused on competition, predation and parasitism (Kingsland 1995).

Part of the reason for the underrepresentation of mutualism in population theory and community ecology is the widespread use of classic Lotka–Volterra type models, which were developed for antagonistic associations and appear to give 'silly' results when the feedback is positive rather than negative. However, these models are now more elaborate and can provide an insight into a continuum of interactions ranging from antagonism to mutualism. Nagging questions, however, remain. What role do mutualistic interactions play in shaping life-history traits of partners? In what way do temporarily positive interactions affect larger ecological entities, such as populations, the organization of communities or ecosystems? What are the long-term positive and negative effects on the fitness of each partner? Are mutualistic interactions particularly success-ful and do they serve as radiation platforms for those species that successfully manage to cope with the aggressiveness of ants? We try to find answers to these questions for associations between ants and their insect partners.

Equipped with the theory of evolution by natural selection, biologists found it easier to think about predation, parasitism and competition (antagonistic interactions), which were crucial for the development of modern ecological thinking. A question immediately arising from this point of view is: How can mutualism exist in the face of exploiters? If positive interactions between the partners of different species lead to an increase in the abundance of each partner or its carrying capacity, then the now more abundant mutualists should be highly attractive to exploiters. Again, one would expect benefits to vary with the population size or density of the partners, because a larger resource of mutualists is more rewarding if it can be exploited. The real world provides persuasive evidence that mutualism does exist and that the net outcome of these interactions is positive at least somewhere in time and space. For example, from the often highly specialized array of pollinators that receive nectar on transferring pollen between flowers, to the myriad of nutri-tional symbionts that fix nitrogen and/or help digestion, positive interactions between organisms belonging to different kingdoms are abundant. Yet, con-temporary textbooks of ecology largely neglect mutualism (Keddy 1990) and in particular, do not ask how this type of interspecific interaction developed and is maintained. This is probably not because mutualism is just another form of exploitation and can be treated under competition, but because it is extre-mely difficult to understand and explain the mechanisms that are simulta-neously at work and eventually lead to different outcomes of species interactions. For example, even though the mycorrhizal associations between fungi and plants are often viewed as mutualistic it is less clear how these interactions affect reproduction and survival of individual plants if fungi invade the root system of different plant species at the same time. Fungi might receive different amounts of nutrients from different plants at different times and actually increase competition at the level of the primary producers. There is only one way to resolve the inherent complexity involved in mutual-isms. In order to understand the ecological role of mutualism in determining the net fitness effects and ultimately the effects on other levels of ecological organization it is important to determine the costs and benefits for each partner over their entire life cycles.

Mutualisms involve ecological and evolutionary processes occurring at scales ranging from individuals to ecosystems. For example, an ant worker that carries home a seed from a plant gains energy, which might contribute to colony survival. For the plant this means that one of its seeds is lost. However, seeds found by ants may subsequently be lost by them and germinate in the vicinity of an ants' nest where herbivore pressure might be reduced and nutrient supply enhanced. The critical question is what proportion of seeds needs to be eaten by ants before such a relationship shifts to a negative/ predatory relationship? Alternatively, one may ask how many seeds must be positively affected by ants in particular environments for dispersal by ants to be at a selective advantage over dispersal by wind or birds. The outcomes of such interactions are probably highly asymmetrical with ants gaining short-term benefits while the benefit for plants might be less obvious if ants even-tually affect plant species composition in an ecosystem. A positive outcome might only be discernible over periods of several years. Similarly, the interac-tion between ants and members of different insect species might lead to immediate benefits for both partners, such as energy-rich food or protection from natural enemies, but it is far less clear whether these interactions are symmetrical and increase long-term fitness of the interacting partners. Equally difficult to discern is how symmetrical the costs and benefits for each partner have to be for mutualistic associations to persist. This is especially relevant for partners with different genetic structures, for example different levels of relatedness between individuals in a population of mutualist 1 compared with mutualist 2, or with life cycles of different lengths. Consequently, rather than defining positive or antagonistic interactions in a typological way we view mutualism and parasitism/predation as two opposites in a continuum of potential outcomes (Fig. 1.1). However, this two-dimensional approach is a gross simplification of the multidimensional, multilevel interactions that occur at various time scales, defined by the life cycles of the partners.

This continuum might be best described along resource gradients, which comprise abiotic conditions, spatial patterns and habitat structure or simply the abundance (density) of each partner. In particular, interactions between ants and their insect partners (herbivorous partners) take place in a multi-dimensional matrix with bottom-up and top-down effects influencing the outcome of interspecific interactions.

The above suggests that mutualisms should be described in terms of resources and therefore interactions are unlikely to be pair-wise. In many cases alternative guild members are equally suitable mutualists. That is, they have direct positive effects on the fitness of their partners, which might vary at least temporally with respect to the net benefit they confer. For a facultative mutualism this can be shown graphically as in Fig. 1.2. In this scenario all three species are able to enter mutualistic relationships and all pair-wise relationships lead to increased

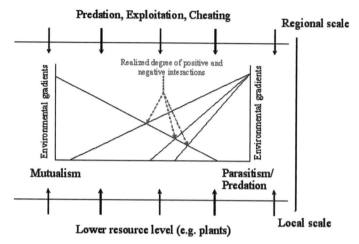

Fig. 1.1. Whether a particular interaction develops into a mutualistic or antagonistic association is likely to depend on environmental gradients such as temperature, moisture, species abundance, spatial structure, density of the partners, etc., which affect the life-history traits and fitness of each partner and ultimately their traits. These processes also operate at a variety of different scales from local to regional, via dispersal. The realized degree of mutualism is a consequence of all the traits contributing to individual fitness constrained by bottom-up (i.e. plant quality) and top-down (i.e. shared predators) effects, which operate at local to regional scales. Below attempts are made to define these axes and identify the mechanisms that connect them.

abundance of both partners at least at some densities. All interactions (e.g. $p_1 - p_8$) are positive; however, they differ in terms of net benefit (left panel: $p_8 < p_5$, right panel: $p_8 = p_5$). If species X becomes unusually abundant then the net benefit of species A and B will be negligible (low values of p_5 and p_8). Alternatively, another conceivable ecological scenario is that species X and species A and B have very different generation times, different generation numbers or modes of reproduction causing time lags in response to increased partner availability. This would suggest that with increasing density of X both partners are competing for access to or services of X (e.g. p_9, p_{10}), especially if their populations are not yet limited by external resources (point a in the lower panel of Fig. 1.2). This scenario can be extended assuming that mutualists A and B are at their maximum densities and limited by resources rather than by access to their partner X. Here mutualist B (e.g. point b in lower panel) might replace mutualist A at higher densities of partner X (point b in the lower panel of Fig. 1.2) (Stanton 2003).

The main message of these verbal arguments is that the ecology and evolution of mutualisms can best be understood in a multitrophic context; that is, if a community ecology perspective is adopted. This includes knowledge of the

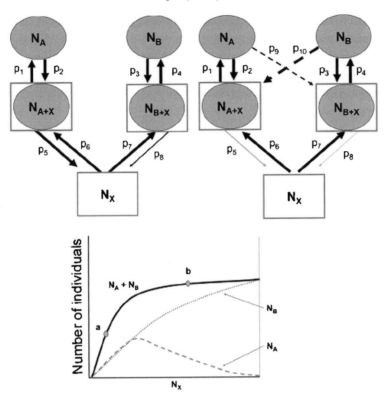

Fig. 1.2. (Top panel) Interactions between individuals of species (X) and two partner species (A, B). N_i is the number of individuals of species i. N_{i+x} represents the number of associations formed between mutualist X and partner species i. Arrows indicate net benefits with the thickness of an arrow representing the net benefit a partner receives from interacting with another. (Upper left panel) Species X receives fewer net benefits from interacting with B than with A. (Upper right panel) Given that the partner species of X negatively affect each other (dashed lines) (p_9, p_{10}) then the net benefit for X will decline. Here the effect of species B is more negative (p_{10}) than the effect of species A (p_9) on the abundance of species B. (Bottom panel) Density dependent effect of population size of species X on the abundance of species A and B. At low densities of species X, A and B increase in numbers in a similar way, while at higher densities of X mutualist B replaces mutualist A. (Modified after Stanton 2003.)

population dynamics, relative fitness costs and gains with different partners, as well as ecological factors and life-history attributes that affect relative population growth rates and, in particular, the abundance of each partner. Depending on the extent to which mutualists are limited by bottom-up and top-down effects, access to partners or competition with other species within a guild, it is expected that interactions can range from strongly positive to negative and that asymmetrical interactions are the rule.

If mutualism is addressed in a community context it can be built upon a solid foundation of ideas and terms well established in community ecology rather then using vague notions like biological markets and trading in commodities. Density dependent competition will frequently affect almost all partners of ants and determine the benefits of interacting with ants. Directly related to this are colonization-competition trade-offs, which link dispersal and the co-occurrence of mutualistic, antagonistic or neutral interactions with ants. The competition–colonization trade-off is an epicentre around which the mutualism–antagonism continuum will be constructed and the effects extended to spatial configurations like those seen in metapopulations.

It is sometimes argued that there is no general theory of mutualism that comes anywhere near the explanatory power of Hamilton's rule, which applies to within-species co-operation (Bronstein 1994b, Herre *et al.* 1999). This is possibly because within-species interactions occur on much finer scales, with much less variation in ecological parameters than in mutualistic interactions, where individuals belong to different species and may interact in many different environments. Nevertheless, the basic idea of Hamilton's rule is that cost–benefit ratios determine whether positive interactions ultimately yield a net positive outcome and are thus likely to develop. The same principle should apply to mutualism. Of course, for interspecific mutualism these costs and benefits are more difficult to measure compared with the inclusive fitness of related individuals, which is defined as the individuals' relative genetic representation in the gene pool of the next generation. The success of this idea rests on the elimination of ecological boundary conditions or spatial configurations of members of the interacting population. Extension of this idea can be found in reproductive skew models, which include those ecological and behavioural interactions that determine how these factors jointly influence the apportionment of reproduction in socially organized colonies (Keller and Chapuisat 1999). Including ecological, spatial and genetic factors in a theory of mutualism is even more important because interactions between different species will typically be influenced by both bottom-up and top-down processes. Nevertheless, the relative magnitude of costs and benefits will determine the success of a particular strategy and thus its genetic representation in the next generation.

The impetus to write this book was the desire to identify and compare conflicts of interests in ants and partners of ants and to understand what factors influence the variation in costs and benefits when interacting in different environments. These variations ultimately determine the outcome of interactions, which might be antagonistic or mutualistic at different times, and conditional on place and partners or their densities. To determine whether there are any general principles or at least some consistent patterns, models and individual case studies are used to test theory. Every attempt is made to avoid getting lost in the fascinating array of highly specialized, coevolved mutualisms and concentrate on the underlying principles and whether they are generally applicable to different partners of ants.

Twenty years ago, in his contribution to the book of Douglas Boucher (1985) on mutualism Daniel Janzen (1985) wrote '... mutualism is not a complex subject and is easily explored through the application of common sense and history knowledge'. He goes on to say: '... mutualism has been thought to death ...' and '... the authors of this volume apparently think that there is something to say [about mutualism], but I wonder if we are not beating a dead horse'. Over the past 20 years there have been many more case studies. The study of mutualism is livelier than previously and is now mainly concerned with finding general patterns and developing a broad-based theory to account for these patterns. There is little doubt that eventually a better understanding of the forces that govern mutualistic interactions will make a significant contribution to modern ecological thought.

Mutualistic interactions occur widely between different groups of organisms, making it impossible to cover the whole field. As indicated above this book is restricted to mutualism between ants and their insect partners (mainly lycaenids and homoptera) because these insect groups have been particularly well studied. We start with an historical perspective of mutualism (Chapter 2) and then discuss different theoretical approaches to two-species mutualistic interactions (Chapter 3). Then the emerging patterns in ant–myrmecophile interactions are addressed, blending information from major partners of ants (Chapter 4). Next, details are presented of the associations that evolved between aphids and ants, for which considerable information is available (Chapter 5). As indicated above, mutualism is not only interesting in terms of the way the two partners interact but also in the way the effects of these interactions extend to higher levels of organization, such as communities and ultimately ecosystems. The aim of the next chapter is to present an integrated account of the processes and emerging patterns associated with mutualistic interactions between ants and their partners at different levels of organization (Chapter 6). Because mutualisms between ants and their partners are multilevel issues we finally focus on a hierarchical perspective integrating key points from the life-history level to applied problems at the ecosystem level, including the invasion of exotic species of ants and their subsequent effects on community structure (Chapter 7). We end by pinpointing frontiers in research on mutualism involving ants and their insect partners.

IN SUMMARY, the study of the range of mutualistic interactions between ants and their partners requires a resource-based cost–benefit perspective. Whether the outcome of such an interaction is a predator–prey or mutualistic one is dependent on what each partner has to offer relative to the needs of the other. Because of its multitrophic nature, mutualism is firmly based in ecology and deals with issues such as density dependence, colonization–competition trade-offs, bottom-up and top-down forces, time lags and fitness costs and benefits.

2

Historical perspective

Compared with competition and exploitation (predation and parasitism) mutualism has been very little studied by field and theoretical ecologists. In 1986 May and Seger recorded that the ratio of papers on competition:exploitation:mutualism published in ecological journals was 4:4:1 and in terms of pages devoted to these subjects in ecological textbooks it was 5:6:1, and this marked prevalence of publications and pages in textbooks on competition and exploitation over that devoted to mutualism still prevails (Fig. 2.1).

This is surprising since a vast number of mutualistic relationships, many of which are still incompletely described, are known from nature, for example: pollinating insects, symbiotic micro-algae in corals etc., and symbiotic nitrogen-fixing bacteria. Besides these, which are fundamental for the functioning of organisms, there are many more exotic examples, for example cleaning symbioses of coral fish, associations of hermit crabs with sea anemones and luminescent bacteria with fish and cephalopods.

It is suggested that the gender bias in science may have encouraged a male view of life as a contest, 'like a football game' (Diamond 1978). However, there has not been a noticeable change in emphasis in this respect associated with the dramatic change in the sex ratio among biologists, as least in western societies, which has occurred since 1978. More recently Turchin (2003) suggested that more attention is devoted to competition and exploitation than mutualism because such interactions appear to be more important for population dynamics. Indeed the consumer–resource dichotomy has been central to understanding predator–prey interactions and competition (e.g. MacArthur 1972, Murdoch *et al.* 2003, Turchin 2003). However, this view is challenged by Holland *et al.* (2005) who argue that consumer–resource interactions are central to nearly all mutualisms.

Darwin (1890) was certainly aware of mutualism and in particular of the relationship between certain aphids and ants as the following quotation indicates.

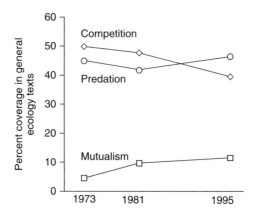

Fig. 2.1. Percentage coverage of competition, predation and mutualism in general ecological texts from 1971 to 1999. (After Holland *et al.* 2005.)

'One of the strongest instances of an animal apparently performing an action for the sole good of another, with which I am acquainted, is that of aphides voluntarily yielding, as was first observed by Huber, their sweet excretion to ants: that they do so voluntarily, the following facts show. I removed all the ants from a group of aphides on a dock plant, and prevented their attendance during several hours. After this interval, I felt sure that the aphides would want to excrete. I watched them for some time through a lens, but not one excreted; I then tickled and stroked them with a hair in the same manner, as well as I could, as the ants do with their antennae; but not one excreted. Afterwards I allowed an ant to visit them, and it immediately seemed, by its eager way of running about, to be well aware what a rich flock it had discovered; it then began to play with its antennae on the abdomen first of one aphis and then of another; and each, as soon as it felt the antennae, immediately lifted up its abdomen and excreted a limpid drop of sweet juice, which was eagerly devoured by the ant. Even quite young aphides behaved in this manner, showing that the action was instinctive, and not the result of experience. It is certain, from the observations of Huber, that the aphides show no dislike to ants: if the latter be not present they are at least compelled to eject their excretion. But as the excretion is extremely viscid, it is no doubt a convenience to the aphides to have it removed; therefore probably they do not excrete solely for the good of ants. Although there is no evidence that any animal performs an action for the exclusive good of another species, yet each tries to take advantage of the instincts of others, as each takes advantage of the weaker bodily structure of other species. So again certain instincts cannot be considered as absolutely perfect; but as details of this and other such points are not indispensable, they may be passed over'.

Interestingly this quotation comes not from chapter 3 on 'Struggle for existence', but chapter 8 on 'Instinct', which considers the cases that fit ill with his theory of natural selection. Darwin concludes that no instinct can be shown to have been produced for the good of other animals; however, animals take advantage of the instincts of others, such as the production of honeydew.

When he considers the structure of insect societies Darwin believes that the difficulties regarding the sterility of certain members and their structural diversity largely disappear when it is remembered that selection may operate at the family as well as at the individual level. In addition Darwin's 'struggle for existence' is an abstract metaphor as he applies it to carnivores struggling with each other to obtain food to survive, a plant's struggle with a desert against drought and mistletoe struggling with other fruit-bearing plants in order to tempt birds to devour and thus disseminate its seed. Nevertheless, Darwin supports his argument by a plethora of bloody battles that Tennyson aptly capsulated in 'Nature, red in tooth and claw'. That is, Darwin attempted to address the role of mutualism in evolution but left it ill defined compared with competition and exploitation, which are well supported by well-documented and clear case studies. Thomas Henry Huxley (1888) championed the gladiatorial view of natural selection and this view has continued to dominate the thinking and writings of western scientists right up to the present day.

The gladiatorial aspect of Darwin's view of the world did not receive universal acclaim. Count Petr Kropotkin, a Russian emigré anarchist who was living in exile in the UK at this time, responded to Huxley's publications by proposing that the struggle for existence frequently resulted in co-operation and mutual aid rather than combat. In his book entitled *Mutual Aid* (Kropotkin 1902) he presents the views of the Russian School of Darwinian critics who rejected Malthus's claim that competition, in the gladiatorial mode, must dominate in an ever more crowded world. Although accepting that this struggle can lead to competition for personal benefit Kropotkin emphasizes the struggle organisms have against the harshness of surrounding physical environments rather than against other members of the same species. Their struggle against the environment is best waged by co-operation among members of the same species – by mutual aid. That is, Kropotkin created a dichotomy within the general notion of struggle. Organisms struggle against organisms of the same species for limited resources, which leads to competition, or they struggle against the environment, which leads to co-operation. Although Darwin acknowledged that both existed he emphasized the competition aspect and his disciples championed 'nature, red in tooth and claw'. Kropotkin did not deny this aspect of the struggle but argued that the co-operation aspect was possibly of equal or greater importance in nature as a whole.

Darwin and Kropotkin gained their experience of nature in similar ways but in different parts of the world: Darwin during the voyage of the *Beagle* (1831–36) and Kropotkin during the five years (1862–67) he spent in Siberia shortly after Darwin published *The Origin of Species* (1859). Like Darwin

Kropotkin went as a 'gentleman observer', but to a military expedition. He traversed over 50 000 miles and had ample opportunity to study in detail the geology, geography and zoology. However, he travelled in sparsely populated areas where frequent catastrophes threatened the few species able to survive there. As a potential disciple of Darwin Kropotkin looked for evidence of the competition that reading *The Origin of Species* had led him to expect. In a letter to his brother he wrote: 'vainly looked for the keen competition between animals of the same species, which the reading of Darwin's work had prepared us to expect but were struck instead by the many adaptations for struggling, very often in common, against the adverse circumstances of climate, or against various enemies'. In addition, for Kropotkin and his fellow biologists Malthus's argument that inexorable increase inevitably strained potential supplies of food and space was distant from the Russian reality – huge land mass dwarfed by its sparse populations.

When in England, Petr Kropotkin commented on the striking difference between the zoologists in his native Russia and his adopted England. Like Kropotkin they saw a great deal of mutual aid where Darwin and Wallace saw only struggle. Kropotkin attributed this in part to the Malthusian ethos in England but also emphasized another factor. Russian zoologists investigated enormous continental regions in the temperate zone, where the struggle of the species against natural conditions is more obvious, whereas Wallace and Darwin primarily studied the coastal zones of tropical islands, where over-crowding is more noticeable.

Thus, the two Englishmen who simultaneously developed the selection theory shared two experiences: a voyage to the tropical rainforests of the Equator and a sympathetic reading of Malthus' *An essay on the principle of population*. Most Russian evolutionists shared two experiences that were roughly opposite to these: travels upon a vast continental plain (with sharply contrasting and swiftly changing environmental conditions) and an aversion to the views of Malthus (Todes 1987, Gould 1988).

So, although the role of mutualism was actively debated early in the development of evolutionary theory it did not attract the attention and research activity subsequently devoted to competition, parasitism and predation. The most likely reason for this is not a lack of awareness of its possible importance but the difficulty of defining and measuring the costs and benefits of the interactions in mutualistic relationships. However, the situation is changing and there are indications that mutualism is beginning to attract a level of theoretical and experimental studies comparable to those associated with the dramatic post-war increase in understanding of competition, parasitism and predation.

IN SUMMARY, mutualism has been very little studied compared with competition and exploitation. Although aware of mutualism Darwin mainly supported his theory with examples of the 'struggle for existence'. In contrast, Kropotkin, although very supportive of Darwin's theory, claimed that co-operation was possibly more important. It is likely, however, that the difficulty of defining and measuring the benefits and costs, rather than an unawareness of its importance, has been the main factor hindering studies on mutualism.

3

Theories on mutualism

3.1 Theories on co-operation

The theory of co-operation between kin and closely related individuals has flourished since the publication of Hamilton's $C/B < r$ rule, stating that the relatedness (r) of an individual that profits from a co-operative (altruistic) act must be higher than the cost (C)/benefit (B) ratio this act imposes. This inclusive fitness concept is best explained by a simple example. Consider a pair of diploid brothers ($r = 0.5$) who share, on average, 50% of their genes. If one of them sacrifices his own fitness by not reproducing ($C = 1$) but helps his brother to rear his offspring successfully the following condition must be fulfilled. In order for C/B to become smaller than r the benefit for the receiver of the altruistic act must at least double before the altruist will gain representation in the next generation. Evidently, to beat the disadvantage of not reproducing when a high coefficient of relatedness is involved, low costs or large benefits are needed. As the benefit of co-operation decreases rapidly with declining relatedness, it becomes clear that the ability to discriminate between related and unrelated individuals is vital for the evolution of co-operation. Given that ants and partners of ants like aphids, lycaenids or coccids are often socially organized, relatedness is an important issue also in the co-operation between members of different species (mutualism). Therefore, it might be helpful to briefly explore the major theories and mechanisms involving co-operation/mutualism trajectories (Table 3.1). Essential for all these theories is that there are costs and benefits involved in entering co-operative or mutualistic interactions and it is the relative magnitude of costs and benefits that determines the outcome of the interaction. The theories differ substantially, however, in the way they incorporate genetic, life-history, population or environmental information.

The explanation of co-operation in the face of cheating is a particular problem for Charles Darwin's theory of natural selection, which emphasizes

Table 3.1. *General theories and mechanisms of beneficial interactions among unrelated individuals arranged in a co-operation–mutualism gradient and with increasing applicability to mutualism*

Game theory and retaliation
Definition: Punishment or refusing future interactions with a partner after an act of cheating.
Example: Predator inspection. Scorekeeping is necessary. No example for insects.
Theory: Trivers (1971), Axelrod and Hamilton (1981).

Spatial games or neighbourhood interactions
Definition: Interactions with only nearest neighbours, selection against cheaters when abundant.
Example: Aphids and ants.
Theory: Brauchli *et al.* (1999), Nowak and May (1992), Doebeli and Knowlton (1998), Addicott *et al.* (1987), Antolin and Addicott (1991).

Trait group selection
Definition: Demes with many co-operative alleles have higher fitness than those with few (differential trait group productivity).
Example: Nest building behaviour of socially organized lycaenids and differential degrees of association with ants (Ruf *et al.* 2003); unrelated, but co-operatively nest founding ant queens (Rissing *et al.* 1989).
Theory: Sober and Wilson (1998), Dugatkin and Mesterton-Gibbons (1996), Dugatkin and Reeve (1994), Dugatkin (2002).

By-product mutualism
Definition: Selfish activity by one partner inadvertently creates an indirect benefit for the other partner that outweighs the cost of the selfish act.
Example: Ants tending homopterans, coccids, plants (Way 1963, Pierce *et al.* 2002); ants attending homopterans also provide some protection for plants against herbivores (Beattie 1985).
Theory: Brown (1983), Connor (1986, 1995).

Market effects
Definition: Decisions to co-operate and with whom to co-operate are based on a comparison of the potential benefits received by different potential partners. Supply–demand perspective.
Example: Yucca–yucca-moth interaction.
Theory: Noe and Hammerstein (1994), Hoeksema and Bruna (2000), Hoeksema and Schwartz (2002).

Competition-colonization trade-off
Definition: For each mutualist the other is a resource with parasites/cheaters as competitors for that resource. Differences in the ability to colonize new habitats and competition for resources will determine the outcome of interactions; that is, mutualism is best understood using theories of species coexistence. The emphasis is on trade-offs, density dependence and spatial aspects.
Example: Ant–plant interactions; ant–insect interactions.
Theory: Levins and Culver (1971), Yu (2001), Abrams and Wilson (2004).

Source: Modified after Yu (2001).

the survival of the fittest. Many insect societies, such as ants, termites, bees and wasps are perceived as well-organized groups dominated by altruistic individuals engaged in peaceful co-operation. However, life within a colony is not always as harmonious as it might appear (Keller and Chapuisat 1999). Understanding how potential conflicts among selfish but related individuals are resolved is of primary importance for understanding the evolution of co-operation at the colony level. Given its simplicity it is somewhat surprising that kin selection theory successfully predicts the evolution of co-operation and partitioning of reproduction in real world scenarios. For example, Hamilton's rule does not explicitly take into account social interactions or changing ecological settings. Recent extensions of Hamilton's rule, like reproductive skew models (Keller and Reeve 1994, Heinze 1995) include ecological, genetic and social factors in a single explanatory framework and aim to determine how these factors jointly influence the realized degree of reproduction of colony members. Clearly, there is now a dynamic view of the equilibrium between co-operation and conflict in within-species systems.

Game theory also primarily deals with intraspecific co-operation, though between non-kin. The simplest game is a symmetric, two-player, two-strategy game (labelled Prisoner's Dilemma). Axelrod and Hamilton (1981) showed that a strategy called Tit-For-Tat yields the highest pay-off in terms of fitness and is resistant to invasion by pure cheaters. Tit-For-Tat implies co-operation in the first round and copying the behaviour of the other player (either co-operate or cheat). Thus, co-operation breeds co-operation and cheating breeds cheating and this strategy proved to be superior to pure co-operators or cheaters. The co-operator's dilemma shows the difficulties of achieving co-operation benefits among a group of unrelated individuals. There are literally thousands of variants of the basic Tit-For-Tat strategy (Nowak *et al.* 1994) depending, for example, on the implementation of the history of interaction (memory capacity) and how it is extended in space and time. For retaliation to work in the case of cheating each partner must recognize the other player individually, which limits the application of this theory to organisms with considerable cognitive capacities. Both ants and their insect partners lack this kind of intelligence. In addition, retaliation strategies are likely to fail if partners are very mobile, live in large groups or if conditions for repeated interactions change very quickly due to environmental influences. Most likely all interactions between insects and ants have these characteristics.

Two assumptions underlying these analyses are inconsistent with the biology of most interspecific mutualisms: (1) players compete directly with their partners and thus are able to keep a score of the pay-offs associated with co-operation, and (2) defection and cheating are constant. Two noteworthy and important extensions of simple symmetric Prisoner's Dilemma settings are provided by Nowak and May (1992) and Nowak *et al.* (1994), who include space, which is the interaction with immediate neighbours, and by Doebeli and Knowlton (1998), who varied pay-offs according to the investment. They assume that 'healthy organisms have more to offer to their partners' and each partner constantly evaluates initial offers and increases or decreases investment in response to past pay-offs. Because of differences in investment and rewards this behaviour might lead to local differences in the costs and benefits of the interaction. Patchiness, that is, limited dispersal abilities with no cognitive skills, and 'strategies' in a two-dimensional lattice generate chaotically changing spatial patterns, in which co-operators and defectors persist indefinitely. Similarly, for mutualism to evolve an increased investment in a partner must yield an increase in returns, with the spatial structure modulating the costs and benefits. Although more biological realism is now incorporated in these modified Prisoner's Dilemma models, which allow continuous variation in investment and pay-offs over the course of a game (Killingback *et al.* 1999), there is no experimental evidence that mutualistic systems operate similarly. Probably one of the greatest disadvantages of Prisoner's Dilemma models is that they do not incorporate fluctuations in population size with changing pay-offs and disregard the associated life-history trade-offs.

Trait groups are defined as populations that are reproductively isolated from other groups and each individual of a particular group interacts primarily with other individuals of that group, which might be temporarily and spatially isolated from other groups (Wilson 1975, 1983). Thus, the level of selection is divided between individual traits and group traits and fitness is a composite of within- and between-trait group fitness. In the trait group model, trait groups are embedded within a larger interbreeding population incorporating a basic notion of metapopulations. Selfish individuals may increase in frequency within groups, but trait groups with many co-operators are thought to produce more offspring than those with few. After reproduction, dispersal might occur and both selfish and altruistic copies are exported. Eventually trait groups may be formed again during the life cycle, e.g. when co-operating individuals preferentially associate again. Co-operation or altruism evolves under this scenario when the increased productivity of groups with many altruists outweighs the within-group advantage cheaters would receive (Dugatkin and Reeve 1994). Thus, a prerequisite for this mechanism to work is differential productivity of trait groups. Does this model apply to interactions between ants and their partners? Can individuals of local demes, which act altruistically and pay the price of attendance when entering associations with ants, be more successful than selfish individuals of an ant-adverse trait group where the individuals do not incur the direct and indirect costs associated with ant attendance? It is relatively easy to imagine situations where the protection service provided by ants benefits all individuals that pay the costs of attendance and might do better than unattended colonies (demes). A critical aspect is the reformation of trait groups with purely altruistic features after dispersal. Recent experiments clearly show that partners of ants adjust the level of honeydew/nectar production to the level of ant attendance and try to reduce individual costs. In addition, the trait group model fails to incorporate population attributes and cannot explain why closely related species living in the same environment might show very different degrees of co-operative behaviour. There appear to be just two possible outcomes: either co-operation or defection representing mutualistic relationships or no relationship. The large majority of partners of ants, however, maintain unspecific and opportunistic (facultative) associations with ants that are not explained by this model. There is no example that conforms to this scenario. That is, there is no evidence that mutualistic associations subdivide differently into isolated demes (trait groups) than non-mutualistic associations.

By-product mutualisms arise because each animal must perform some kind of behaviour that may benefit another individual as a by-product (Brown 1983). Although originally coined for intraspecific co-operation it might equally well apply to interspecific co-operation. In order to understand how by-product mutualism operates it is helpful to introduce different types of environment. Harsh environments are characterized by, for example, many predators, little food, adverse weather conditions or any factor that might reduce individual fitness. Mild environments are the converse. Harsh and favourable habitats need not be separated physically if time is a relevant factor, because the same environment can be favourable in one period and harsh in another, due to seasonal changes. The underlying hypothesis is that co-operative behaviour is an incidental consequence of ordinary selfish behaviour but only in adverse environments (Mesterton-Gibbons and Dugatkin 1992). Any decision an animal makes is subject to what is called a *boomerang factor*: a certain probability that a cheating or non-co-operating individual will fall victim to its own cheating and suffer fitness costs. The typical interaction is still described by the Prisoner's Dilemma game with pay-off matrices now supplemented with an additional parameter which measures the adversity of the environment. According to this theory, harsh environments are more likely to produce such boomerang effects. In this respect it is similar to Kropotkin's view. Therefore, co-operation via by-product mutualism might be more frequently found in environments with a common enemy or that are adverse. In favourable environments it is more rewarding to cheat or at least not to co-operate because nothing can be so gained. There appear to be a number of examples of the boomerang factor, e.g. co-operative hunting in lions, which appear to co-operate when hunting large prey or in territories of rivals but hunt alone for small prey (Heinsohn and Packer 1995).

There also appear to be a number of examples for ants and their partners that might qualify as by-product mutualisms. For example, honeydew of aphids, coccids and membracids is a waste product, which is an inevitable consequence of feeding on phloem and xylem sap of low nutritional quality. In order to survive on this plant resource they have to process large quantities of sap to extract sufficient nitrogen, which is mostly available only in low concentrations. As a consequence, ants that collect the energy-rich excreta provide hygienic services, which are beneficial to their homopteran partners. The necessary excretion of a waste product might thus be seen as a by-product facilitating a mutualistic service. In a harsh environment, with many natural enemies, this service might gain in relative importance because of the addition of another function, which is protection against natural enemies. Thus, in harsh environments a functional switch of an original by-product mutualism might facilitate the development of close (obligate) mutualisms. So the notion that individuals co-operate or form mutualistic associations when it is beneficial (e.g. as a by-product), or otherwise forgo this option, requires few assumptions and is an integral part of evolutionary thinking. Many mutualisms might be of the by-product type and require few assumptions in a cost–benefit context.

Biological market models are based on economic theory and assume that mutualisms are systems of mutual exploitation (Herre *et al.* 1999) developing along a continuum from obligate mutualism to parasitism (Hoeksema and Schwartz 2002). In this case species interactions operate as 'biological markets' in which both partners trade commodities or services, which they can produce at little or no cost and obtain benefits that they cannot produce (Noe and Hammerstein 1994, 1995). If mutualisms operate as biological markets costs and benefits must be expressed in demographic and fitness currencies. The magnitude of mutualistic effects may depend on extrinsic factors, which make it difficult to measure net gains. A gradient of resource availability is the driving force that mediates investment to receive or provide specific commodities or services. Market models, however, have a limited application in biology. For example, most animals do not have the cognitive capabilities to remember who meets whom, what the traders know or how to maximize fitness using a price-setting process. In real markets, once prices are set and contracts are agreed they are enforceable at little cost and, therefore, there is little incentive for cheating because no side profits. In addition, a major difficulty is translating single market interactions into costs and benefits in the context of reproduction, survival and population-level effects. Therefore, the application of market theory to biological problems is likely to be limited. Furthermore, fitness is a composite characteristic including interactions in different environments and with different partners, but market models trade only one commodity. In addition, the precise mechanisms by which partners agree a particular price remains obscure. Progress is dependent on budgeting specific costs and benefits and determining how they are related to the fitness of the organisms.

Finally, the *competition–colonization theory* (Levins and Culver 1971, Yu and Wilson 2001) uses a resource framework to explain the existence of mutualism in the face of exploitation and cheating. It incorporates competition and colonization trade-offs and places mutualism in the well-defined field of species coexistence, using mechanisms of population regulation (Yu 2001, Yu and Wilson 2001). This has been achieved by adopting a more mechanistic approach, which has already proved successful in the analysis of species coexistence, using density dependent population growth and mortality rates, spatial refuges, spatial heterogeneity, colonization abilities and trade-offs. When applied to mutualistic/ antagonistic interactions, the theory assumes that different species can exist along a mutualist–antagonist continuum by specializing on different subsets of the resource spectrum of the partner. The ability to migrate is a necessary precondition to arrive at these preferred resources. Which of the potential outcomes will prevail depends on several factors, including, but certainly not limited to, the strength of the trade-off, which is influenced by the relative need of the partner for a resource, life-history considerations, and spatial heterogeneity influencing colonization and extinction rates in a metapopulation context. Although this theory has mainly been used to account for plant pollination systems, such as the figs–fig moth, or ant–plant relationships, it is general enough to provide an insight into the mechanisms allowing the coexistence of partners of ants with very different degrees of myrmecophily. For example, four species of ants belonging to two genera (*Allomerus, Azteca*) live on a single species of ant plant (*Cordia nodosa*) (Yu *et al.* 2001). The relative colonization abilities of the different ant species appear to be a function of plant density, with the relative abundance of the *Azteca* species declining with increasing host plant density while the reverse is true for *Allomera* species. *Azteca* queens are better long-distance migrants and *Allomerus* colonies are more fecund. Thus, the main mechanism for species coexistence in this system is thought to be the spatial heterogeneity in plant densities and colonization– fecundity trade-offs. More examples will be provided and described in greater detail in Section 6.5; in particular, the largely pair-wise view of these interactions will be extended to multispecies mutualisms.

In view of the variety of mutualistic interactions, of which only a subset is described for ant and ant–partner interactions, it is reasonable to assume that no one theoretical approach is particularly suitable as each has its strengths and weaknesses. However, those theories that place mutualisms along a continuum from mutualism to antagonism, with mechanisms operating along these gradients and at different levels of organization, hold considerable promise of providing an insight into the ecology and evolution of positive and negative interspecific interactions. Some of these models are described below.

3.2 The predictions

Understanding the origin and ecological forces necessary to maintain beneficial interactions with one or several partners is a major challenge, especially if the benefits are difficult to measure or very small when measured over short periods. Contrary to negative interactions between two species, which appear to be intuitively more vulnerable to immediate population decline and extinction, positive feedback is bound to result quickly in overpopulation unless there are mechanisms that limit population growth (Nicholson 1933). In a multispecies world this would quickly lead to instability and ecosystems with more species would tend to be more unstable (May 1972). As a consequence, mutualistic relationships continue to be seen as difficult to understand and the theory advances much slower than in other areas of community ecology, such as intra- and interspecific competition or predator–prey/host–parasitoid interactions. Recently, however, positive interactions have attracted attention; in particular, the view that mutualistic and antagonistic interactions are two endpoints along a continuum rather than different types of interaction has prevailed and many stabilizing mechanisms have been suggested.

There is a great variety of models (Hoeksema and Bruna 2000), many of which can be associated with particular mutualistic interactions and some are even specific for obligate or facultative relationships. They are based on game theory or biological market, Lotka–Volterra, life-history or metapopulation models. For example, Bronstein (1994a) makes three general predictions about mutualistic interactions and argues that (1) the fitness costs of facultative mutualisms are less dependent on the presence of the partner than obligate mutualisms, (2) if the interactions between two partners are affected by the abundance or behaviour of a third party then the outcome of these interactions might vary greatly and (3) benefits for one partner may vary with the abundance of the other (density dependence). The benefits for partners of ants, especially, might be dependent on the number of ants recruited (Harmon and Andow 2007), which might change as the population increases in the course of the interaction. Recent models incorporating a mutualism–parasitism continuum result in the general predictions discussed in the next section. Obviously, obligate mutualistic relationships must include mechanisms or carrying capacities that limit the growth of one or both partners. These mechanisms are explored in more depth in later sections. A further general suggestion is that highly specialized one-to one obligate mutualisms should be rare and limited to species-poor communities (Williams 1966, Howe 1984, Hoeksema and Bruna 2000). This is because they are only likely to be realized in a narrowly defined environment, which probably limits the abundance of at least one partner.

3.3 Models

Can models be used to help us understand positive and negative interactions between organisms? The short answer is that they help one to focus on the main problem, but there are also a number of other reasons for starting with a theoretical construct. A model is an abstraction of the real world and includes mechanisms that may be involved in natural processes (Fig. 3.1). As a consequence, the degree of abstraction that is necessary or possible is critical. In developing a model no attempt is made to incorporate every natural process, otherwise it becomes as difficult to understand as nature. Models are caricatures of nature and arguing that they do not accurately represent nature misses the point. The question is not whether a model is good or bad but whether its assumptions are useful in understanding a particular process.

In order to evaluate the usefulness and predictions of a model one has to be clear what assumptions have been made. These assumptions are about the nature of natural processes and need to be precisely formulated so that direct relationships can be expressed in equations. The logical qualitative and quantitative consequences are the predictions that can be tested experimentally. In many cases models are the only means of making inferences, for example when working with endangered or rare species on which it is difficult or impossible to do experiments, or when interested in the fitness consequences that interactions with ants might have on the population dynamics of the partners of ants.

Experimental ecologists increasingly are aware that the question is not whether a model is correct or false but whether it helps in identifying the mechanisms or determining the patterns in a system. Similarly, modellers seem to be better able to communicate their approach to field biologists, possibly because there are now well-developed tools for presenting the assumptions, transformations and outputs. Even simple spreadsheet programs are now available for doing complex modelling exercises (Donovan and Welden 2001) and visualizing the consequences of particular assumptions.

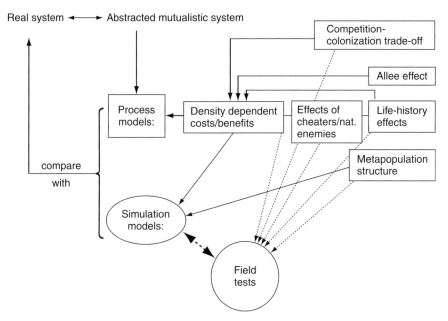

Fig. 3.1. Simplified diagram of the relationship between a natural system (e.g. mutualistic relationships between organisms) and different modelling approaches that incorporate the different mechanisms that are likely to affect the fitness, population dynamics and abundance of interacting mutualists. Ideally, the models should result in field experiments that test the underlying assumptions and predictions. An abstraction of a real system must omit many biological aspects but the resultant simplification should be easier to understand than the real system.

Another important function of models is to generate new ideas about the processes important for a particular relationship, which ideally can then be tested experimentally. For example, density dependence is easily introduced into models of mutualistic interactions, and the outcome in terms of stability of the system tested experimentally (Addicott 1981, Wolin 1985). However, conclusively demonstrating that density dependent processes operate in the field remains a challenge. Models of mutualism with simple processes and behaviour that can be easily followed will mainly be addressed here.

Three general classes of models in particular appear to capture the key features of mutualistic interactions within a consumer-resource dynamic framework using theories of predator–prey interactions, life-history and spatial constraints. These models have a very distinct history and arise from equilibrium and non-equilibrium concepts (DeAngelis and Waterhouse 1987). The specific features of some of these models are addressed and the major assumptions, predictions and suggested mechanisms involved in the mutualistic interactions incorporated in these models are summarized. An emerging general feature of most of these models is that when the abundance of the partners is low the mutualism is more stable than when they are abundant, suggesting a strong density-dependent component in mutualisms.

3.3.1 Lotka–Volterra type models; functional response models

The starting point of many models of mutualism is the basic Lotka–Volterra predator–prey or competitive interaction approach and, as a consequence, they incorporate many of the assumptions implicit in these models of negative inter-specific interactions. For example, as in the logistic growth model, Lotka–Volterra competition models lack the notion of age- or genetic structure, migration or time delays, or monophagous predators. Interactions are deterministic and occur over small spatial scales. In addition, adopting a chemical mass action approach and assuming that the response of a population is proportional to the product of its biomass has been criticized by many ecologists (Kingsland 1995). Adopting this view, resources are finite and competition inevitable. However, the same logic can be applied to positive interactions. The only difference is that the 'competition coefficients, α and β' in the Lotka–Volterra competition model are interpreted as a 'mutualism coefficient' by making their values positive.

In its simplest form the two-species mutualism model is written as:

$$\frac{dN_1}{dt} = r_1 N_1 \left(\frac{K_1 - N_1 + \alpha N_2}{K_1} \right) \tag{3.1}$$

and

$$\frac{dN_2}{dt} = r_2 N_2 \left(\frac{K_2 - N_2 + \beta N_1}{K_2} \right), \tag{3.2}$$

where N_x is the number of individuals, K_x is the carrying capacity and r_x the instantaneous rates of increase of species 1 and 2, respectively. The constants α and β are mutualism coefficients of species 2 and 1, respectively, indicating the effect one species has on the other. For example, when $\alpha < 1$ the effect of species 2 on species 1 is less than the effect of species 1 on its own members. Conversely, when α is > 1 the effect of species 2 on species 1 is greater than that of species 1 on its own members. The traditional way of analysing such differential equations is by studying the behaviour of the zero growth isoclines. The isocline is the set of values of N_1 and N_2, for which one of the two populations neither increases nor decreases. The zero growth isoclines are found by setting equation (3.1) and (3.2) $= 0$:

$$N_1 = K_1 + \alpha N_2, \tag{3.3}$$

$$N_2 = K_2 + \beta N_1. \tag{3.4}$$

Equilibrium is where the two isoclines intersect and it can be shown that this equilibrium is stable if $N_1^* > K_1$, and $N_2^* > K_2$. Vandermeer and Goldberg (2003) have categorized the basic forms of the mutualistic interactions and present eight qualitatively distinct outcomes for facultative and obligate mutualisms predicted by this simple two-species system of differential equations. In this Lotka–Volterra system, most obligate mutualists have larger negative carrying capacities than facultative mutualists (Fig. 3.2). Negative carrying capacities indicate that a population cannot grow in the absence of a partner population, for instance a large negative carrying capacity indicates a more obligate mutualistic species than a small one.

Obviously, not all of these combinations of mutualism coefficients make sense. Combinations that predict ever-increasing population growth (e.g. panels a and g in Fig. 3.2) are impossible because a resource must become limiting sooner or later. If both partners are facultative mutualists, stable coexistence is possible. In these equations obligate mutualists have large negative carrying capacities. This might be counterintuitive at first, but it simply means that the more dependent one partner is on another the more negative is the effect on its carrying capacity. It cannot survive without the mutualist. Obligate mutualisms in which both mutualists have low mutualism coefficients (panel c) are also unlikely to occur in nature, because this means that both species are dependent on each other but provide little benefit. Ultimately they would go extinct. If both species have high mutualism coefficients (panel d) there are two possible outcomes. Either the populations grow indefinitely or the populations go extinct. There is only a narrow range of values that allow stable coexistence. Given that obligate mutualisms between ants and their partners are generally rare this could mean that other factors (not implicit in these equations), such as exploiters of mutualistic associations or spatial constraints, might drive obligate associations to extinction.

Combinations of obligate and facultative mutualists are indicated in panels e–h. For the ant–aphid-partner mutualism this means that ants are always the facultative partner (N_1), while other insects are the obligate myrmecophiles (N_2) with different coefficient combinations. In panel e the obligate partner of ants goes extinct, panel f is similar to d with alternative attractors (one with extinction of obligate partner of ants), g ever-expanding populations and h stable coexistence.

The Lotka–Volterra approach is usually considered to be the baseline model, in which competition and mutualism coefficients are phenomenological entities

(a) Facultative mutualists with
    high mutualism coefficients

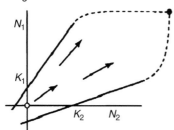

(b) Facultative mutualists with
    low mutualism coefficients

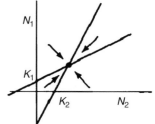

(c) Obligate mutualists with
    low mutualism coefficients

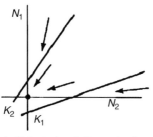

(d) Obligate mutualists with
    high mutualism coefficients

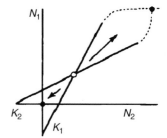

(e) Obligate-facultative mutualists
    with low mutualism coefficients

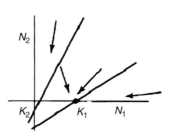

(f) Obligate-facultative mutualists
    with high mutualism coefficients

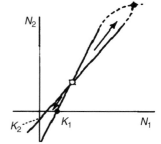

(g) Obligate-facultative mutualists
    with high mutualism coefficients

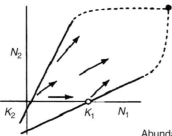

(h) Obligate-facultative mutualists
    with low mutualism coefficients

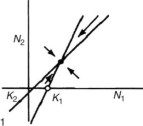

Abundance of mutualist 2

Abundance of mutualist 1

Fig. 3.2. Eight different combinations of facultative, obligate and obligate-facultative partners. Facultative mutualists are able to survive without the partner, but obligate mutualists are not. Stability properties are indicated by the arrows. (After Vandermeer and Goldberg 2003.)

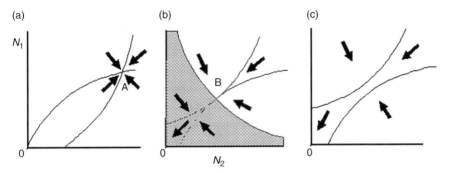

Fig. 3.3. Isoclines for two interacting populations (N_1, N_2). (a) Isoclines intersect at point A, which is a stable equilibrium; (b) touching isoclines lead to extinction in the hatched region or populations are attracted to point B at the edge of the hatched region; (c) isoclines do not meet and populations are unstable. Therefore, only under condition (a) will mutualism exist. (After Dean 1983.)

incorporating many mechanisms at the same time. This, of course, is only a useful first step and should be followed by more mechanistic approaches. Some more advanced examples are given below. The question of interest is: what stabilizes these labile interactions? Is it behaviour, delayed density dependence or spatial heterogeneity? One of the first models that incorporated the idea that benefits of positive interactions might depend on the density of one of the mutualists was suggested by Dean (1983). He used the same basic logistic equations (3.1) and (3.2) and assumed that K_1 and K_2 are not constants but grow asymptotically with the density of a mutualist up to a maximum K_i:

$$k_1 = K_1(1 - \exp(-aN_1 + C_1)/K_1) \tag{3.5}$$

and

$$k_2 = K_2(1 - \exp(-bN_2 + C_2)/K_2). \tag{3.6}$$

K_i is the maximum value of k_i; C_i, a and b are constants. Both populations grow until the density effects limit the growth of N_1 and N_2. As a consequence, the isoclines intersect (stable equilibrium, Fig. 3.3a), touch (Fig. 3.3b, unstable equilibrium), or do not intersect (Fig. 3.3c, no mutualism occurs).

Obviously, whether an equilibrium is stable or unstable depends on the sign and magnitude of the constants C_x, K_x, a, and b. For example, if $C_1 = C_2 = 0$ obligate mutualism will occur if $ab > 1$. Similar to models on host–parasitoid interactions the introduction of time lags (Dean 1983), a functional response in the form of handling time (Wright 1989), competition or predation by a third species (Heithaus *et al.* 1980, Rai *et al.* 1983) and density-dependent interaction functions (Addicott 1981, Wolin and Lawlor 1984) stabilize the basic

Lotka–Volterra type mutualisms. At low densities mutualism is favoured, while at high densities competition is favoured (Ferriere *et al.* 2002, Zhang 2003). Implicit in these results is that complex communities, with interactions ranging from mutualism to competition, might be more stable than previously suggested by May (1972).

A Lotka–Volterra predator–prey model that includes positive and negative effects between species was also used by Neuhauser and Fargione (2004) to explore the shift between mutualism and parasitism. Although originally developed for plant–mycorrhizal interactions the model is general enough to explore the consequences of interactions between ants and their insect partners. Using a predator–prey rather than a competition-based approach is not a major shift in the perspective of interspecific interactions if one accepts that mutualistic interactions are essentially exploitative, with each species using the other to gain benefits (Herre *et al.* 1999). Thus, the principal conclusion is that costs and benefits must be taken into account in order to capture the essence of interspecific interactions. In the predation model a predator might either operate as a predator or a mutualist. The assumptions are that (1) host abundance (mutualist N_1) increases logistically in the absence of predators/parasites/mutualists (here mutualist N_2), (2) N_2 can have positive and negative effects on N_1, with beneficial effects resulting in an increase in the abundance of N_1 while the negative effect is an increase in the death rate of N_1 due to exploitation and (3) density dependent negative feedback processes (self-interference) on N_2 increase its death rate. These assumptions translate into the following equations:

$$\frac{dN_1}{dt} = rN_1\left(1 - \frac{N_1}{K + \gamma N_2}\right) - aN_1N_2, \tag{3.7}$$

$$\frac{dN_2}{dt} = bN_1N_2 - dN_2(1 + eN_2); \tag{3.8}$$

N_1 and N_2 are the densities of the partners, K is the carrying capacity, r and b are the intrinsic rates of increase of N_1 and N_2 respectively, and a represents the exploitation of N_1 by N_2. The parameters d and e describe the density independent and dependent death rates of N_2, and γN_2 quantifies the gain to N_1 of the interaction with N_2. This simple model collapses to the familiar Lotka–Volterra predator–prey model if $e = 0$ and $\gamma = 0$. Using the classic phase plane approach and plotting the isoclines of the interaction between the species gives some interesting results (Fig. 3.4).

A biological scenario that might be relevant for the interactions between ants (N_2) and their partners (N_1) might be as follows. Ants derive nutrients

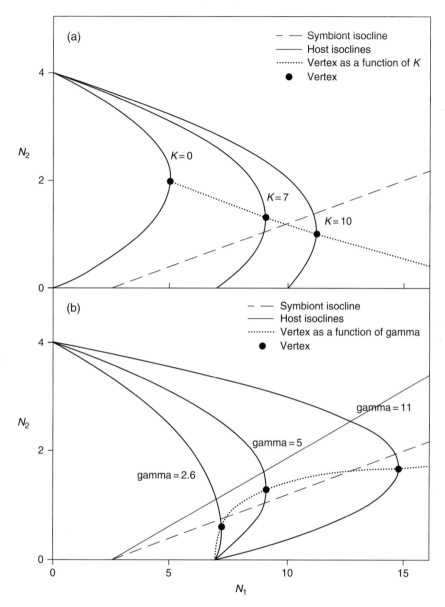

Fig. 3.4. Zero growth isoclines for two interacting populations (N_1, N_2) showing different outcomes of the interactions. Whether an interaction is antagonistic or mutualistic depends upon the relative position of the intersection of the isoclines. If the vertex of the isocline of N_1 is below the intersection then the interaction is parasitic, if it is above then it is mutualistic. (a) Effect of varying the carrying capacity for N_2; (b) effect of varying the interaction benefit γ (gain to N_1 from the interaction with N_2) on the outcome of the interaction. (After Neuhauser and Fargione 2004.)

from their partners and provide them with some degree of protection from
natural enemies. Protection might be particularly advantageous for N_1 if it
lives where the risk of predation is high and especially if there are few alternative energy resources for the ant partner so that tending intensity is high.
From the perspective of the ant, in order to monopolize the energy resource
provided, many workers need to attend the aphid, coccid or membracid
colony. In addition, low numbers of mutualist 1 (N_1) are likely to result in,
for example, allocation of more workers for protection and honeydew/nectar
transport than initially. With increasing numbers of N_1 more resources
become available for the ants and they can focus on those individuals/colonies
with the highest honeydew/nectar output. This incorporates the idea of space.
The effect of the availability of energy on ants can be studied by changing the
carrying capacity K, which increases with energy output of N_1 (Fig. 3.4a).
When all other parameters remain unchanged, the vertex of the zero growth
isocline of the host moves to the right and down, implying that the resource
provided by one partner increases and the interaction can change from mutualistic (vertex of N_1 isocline above the dashed line) to antagonistic (vertex below
dashed line). In general, an increase in resource availability results in interactions changing from being beneficial to being highly costly, at least for one
partner.

In a similar way, the benefits that N_2 provides N_1 can be modified by
changing the value of γ (Fig. 3.4b). The qualitative behaviour of the outcome
of the interaction depends on the slope of the N_2 isocline (e.g. ants). If the slope
is steep (solid line) the vertex of the N_1 isocline (e.g. aphids) is always below the
isocline of N_2, resulting in antagonistic (non-mutualistic) interactions. If the
population growth of N_1 is small (dashed line) an interaction may be antagonistic at small values of γ, mutualistic at intermediate values and antagonistic
again at high values. That is, if the interaction benefit (γ) is low N_1 does not
benefit from the presence of N_2, independent of the population size of both
partners. Intermediate values of γ yield positive effects but surprisingly further
increasing γ can result in antagonistic relationships. This might be because the
increase in population size of N_1 is reached by increasing the numbers (carrying capacity) of N_1 (e.g. protection from natural enemies), which are limited by
the negative effects (Fig. 3.4a) of N_2 on N_1. Thus, with infinitely growing γ, the
model approaches the classical predator–prey model. Similar conclusions were
reached by Zhang (2003) who used the Lotka–Volterra competition model to
introduce density dependent interactions, showing that mutualism is more
likely to occur if densities of the partners are low and exploitation might
prevail if the densities of the partners increase. Here, mutualism is also thought
to promote the competitive ability of a species by increasing its carrying

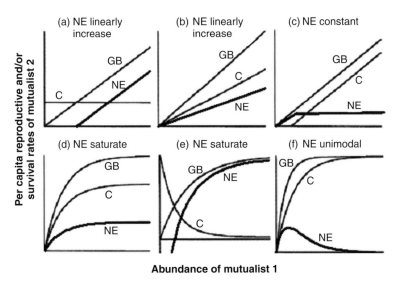

Fig. 3.5. Set of potential functional response curves in terms of reproductive and survival rates of mutualist 2 in response to the population size of mutualist 1. GB, gross benefits, C, costs, NE, net effects. NE = GB − C. (After Holland *et al.* 2002.)

capacity when interacting with the partner species. (Carrying capacity encapsulates the effects of mutualist 1 on mutualist 2 and vice versa.)

Finally, a more explicitly cost–benefit perspective with respect to density dependent effects on mutualism is presented by Holland *et al.* (2002). Contrary to many models on mutualism, which use Lotka–Volterra or type 2 functional response models, that of Holland *et al.* (2002) predicts a more mechanistic functional response reflecting the costs and benefits that might be experienced by interacting partners depending on their population size. Thus, there is still an equilibrium point, but there is a shift towards a more mechanistic integration of life-history variables, which is further explored in the next section. Using this conceptual framework they describe the two-species obligate mutualism between yucca and yucca moths but other mutualisms are equally well described. Some examples of the costs and benefits to one mutualist depending on the population size of the partners are shown in Fig. 3.5.

Benefits to mutualist 2 are zero in the absence of mutualist 1, indicating that if no mutualist is present the curves must start from the origin. This, however, does not mean that mutualist 2 does not experience costs (C) in the absence of mutualist 1, because the association is obligate, implying for example that the investment of mutualist 1 involves a cost if it is not repaid ('association costs'). The shape of the cost–benefit curve depends on whether the costs are fixed (independent of population size of mutualist 1) or variable; that is, if the per capita protection decreases with population size or if mortality rates imposed by mutualist 1 on mutualist 2 increase; for example, ants (N_1) increasingly prey on their nectar or honeydew producing partners (N_2) as the population size of N_1 increases. In Fig. 3.5a costs remain constant independently of the population size of mutualist 1, while the benefits increase linearly with each added mutualist 1 individual. Net benefits, however, only increase when GB > C. In Fig. 3.5b benefits and costs increase with increasing density of mutualist 1 but GB increases faster than C, yielding an ever-increasing net benefit to mutualist 2 with increasing population size of mutualist 1. Ever-increasing net benefits for mutualists, whether there is an initial threshold or not, are unrealistic scenarios. In nature there are many limitations, in addition to the interactions between mutualists, such as carrying capacities, which when exceeded result in a sudden or gradual decline in net benefits with increasing population size of a partner. For example, Fig. 3.5c suggests a situation where the initial cost is negligible up to a threshold level, but C and GB increase in parallel thereafter. Various saturation effects result if C and GB rise asymptotically with the level of benefit conveyed to the mutualist depending on the specific functional response curves (Fig. 3.5d–f). For example, when ants are rare, nectar production by lycaenids or the investment in honeydew of obligatorily myrmecophile aphids and coccids may be more costly than the benefit gained from protection against natural enemies (Fig. 3.5a, e). Similarly, high tending rates of ants might reduce the net benefit to the partners if they have to produce large amounts of honeydew or nectar to appease ants and so avoid being eaten (Fig. 3.5f). Not feeding at an optimal rate might reduce the growth rates of the partners of ants (Stadler and Dixon 1998a). Basically, an infinite number of cost–benefit functions are conceivable with costs probably more often exceeding benefits than indicated here. But as in the previous models mutualism is seen from a cost–benefit perspective and might shift from mutualism to antagonism depending on population size and associated constraints.

In a more formal way the above arguments can be framed in the traditional zero growth isocline analyses of net effects in mutualistic populations. To explore how the population size and dynamics of mutualist 2 varies with the size of the population of mutualist 1 Holland *et al.* (2002) suggest the following model:

$$\frac{dN_2}{dt} = B_n N_2 - dN_2 - gN_2^2. \tag{3.9}$$

N_2 is the population size of mutualist 2, B_n is the net effect on mutualist 2, d is the mortality rate and g is a negative feedback term, which indicates the exploitation of external resources with increase in population size. Essentially, this model indicates that the population grows according to any of the functional responses suggested in Fig. 3.5. This growth is limited by mortality rate and even more so by external constraints. To illustrate the consequences of different functional responses for the population dynamics of a mutualist three different density-dependent functional responses were considered (Holland *et al.* 2002):

(1) the net effect of the positive interaction of mutualist 2 with mutualist 1 increases linearly with the population size of mutualist 1 (Fig. 3.5b),

$$B_n = mN_1 + a, \tag{3.10}$$

(2) the net effects gradually increase and saturate (Fig. 3.5d),

$$B_n = \frac{\gamma_1 N_1}{1 + \gamma_1 N_1}, \text{ and} \tag{3.11}$$

(3) the net effect is a unimodal function of the abundance of mutualist 1 (Fig. 3.5f):

$$B_n = \frac{\gamma_1 N_1}{1 + \gamma_1 N_1} - \frac{\gamma_2 N_1}{1 + \gamma_2 N_1}. \tag{3.12}$$

The parameter γ_1 is the rate at which net benefits are accrued by mutualist 2 as a function of the population size of partner 1 ($\gamma_1 > \gamma_2$). Incorporating these different functional response equations into the population growth equation of N_1 and calculating the zero growth isocline ($dN_2/dt = 0$) results in the following state plane diagrams (Fig. 3.6). The three lines indicate the conditions when the population size of N_2 does not change. Consequently, for a particular population size of N_1, values above the isocline indicate declining and values below the isocline indicate increasing population growth of N_2. When the population size of N_1 is very small the net benefit to mutualist 2 is negative (e.g. Fig. 3.6a), that is there is a minimum density of mutualist 1 above which the interaction becomes positive for N_2. If the net effect on mutualist 2 is of the saturation type or a unimodal function of the population size of N_1, then the zero growth isocline of N_2 increases asymptotically (Fig. 3.6b) or is unimodal (Fig. 3.6c), respectively. The mortality rate and negative feedback components only have an enforcing effect.

Density dependent interactions between mutualists can thus have very different outcomes ranging from negative to positive. Basically, this corresponds to the Allee effect (Allee 1949) for single species populations, where a critical minimum population size is necessary for survival and below which extinction occurs. But as the population grows negative density effects gain in importance as resources are depleted. The above results also suggest that there

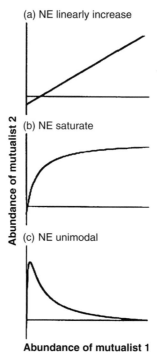

Fig. 3.6. Zero growth isoclines of mutualist 2, for different functional response curves describing the net effects of different population sizes of mutualist 1 on mutualist 2. Values above the isoclines indicate that the population size of N_2 will increase, while values below the isoclines indicate declining population growth of N_2. (After Holland *et al.* 2002.)

might be an optimum population size of the mutualistic partner beyond which the interactions accrue less benefit (e.g. past the peak in the unimodal model, Fig. 3.6c). If these assumptions hold then it is likely that such functions will affect other traits, such as the dispersal strategies of the partners (see models below).

Functional response models provide some insights into and explain why the strength of interspecific interactions can vary depending on costs and benefits experienced by the partners as a result of their changing densities. Understanding exactly how changing densities affect the growth and mortality rates and ultimately fitness of two or more interacting partners in a way that prevents unbounded population growth remains a major challenge. Nevertheless, these models provide a useful framework for exploring the dynamics and stability of population sizes resulting from mutualism.

IN CONCLUSION, in spite of the simplicity of the Lotka–Volterra concept, in which populations are treated as interacting masses rather than individuals, and contrary to the view expressed in many ecological textbooks, simple adaptations of this model have provided useful insights into the forces structuring communities, species coexistence and the ecological dynamics of the mutualism–antagonism continuum. This framework is easily extended to more than two-species interactions. Clearly, these models indicate that a more explicit recognition of density dependent costs and benefits inherent in mutualistic interactions results in a better mechanistic understanding of the diversity of interactions along the mutualism–antagonism continuum. This topic will be considered again in a later section.

3.3.2 Life-history models

Although the above model provides a valuable insight into how overall costs and benefits are affected by the density of a mutualistic partner it is not quite clear how these densities affect actual per capita birth and death rates and in what way costs and benefits feed back into density dependent and density independent effects on mutualism. In addition, facultative mutualistic relationships were not considered. A good starting point for determining these effects in more mechanistic detail are the models of Addicott (1981) and Wolin and Lawlor (1984). Rather than simply affecting the equilibrium density (cf. Holland *et al.* 2002) mutualists can increase r, the intrinsic rate of increase, or both r and the equilibrium density by increasing for example K, the carrying capacity. Simulations demonstrate that it is only mutualistic associations that increase the intrinsic rate of increase, and enhance stability (Addicott 1981). The idea is that when no mutualist is present the per capita birth rate decreases, and the per capita death rate increases as a function of density (Addicott 1981). In the simplest formula with linear relationships this can be expressed in the following way:

$$b = b_0 - aN_1, \tag{3.13}$$

$$d = d_0 + cN_1, \tag{3.14}$$

where b and d are the per capita birth and death rates, b_0 the birth rate at zero density, d_0 the death rate at zero density, which is achieved under ideal (uncrowded) conditions (Wolin and Lawlor 1984). The constants a and c measure the strengths of the per capita density dependent regulatory factors. The larger a is, the more sharply the birth rate drops with each individual added to the population. If there is no density dependence, then $a = 0$ and the birth rate equals b_0, regardless of population size. The same reasoning can be applied to the effect of density on the death rate. Death rate, however, is expected to increase as the population grows. Density dependent and density independent effects can thus be easily incorporated into a logistic growth model $(dN_1/dt = rN_1(1 - N_1)/K$, in which $r = b_0 - d_0$ and $K = r/(a + c))$. Exponential growth is thus only a special case of the logistic model, for example if $a = 0$, $c = 0$ or if N_1 is small relative to the carrying capacity of the population. Mutualism is incorporated by adding density dependent or density independent increases in birth rates. The density dependent birth and death rates in the logistic model are represented by the dashed lines in Fig. 3.7. This is the reference model (null model), with which all others are compared.

Several scenarios can be distinguished.

Density independent effects of mutualism. In this case mutualist N_2 has the same per capita benefit on N_1 at all densities (Fig. 3.7a). If only the birth rate is affected by the mutualist, this can be expressed in the following way:

$$b = b_0 - aN_1 + mN_2, \tag{3.15}$$

$$d = d_0 + cN_1, \tag{3.16}$$

where m is the per capita effect of N_2 on the birth rate of mutualist N_1. That is, the birth rate will increase over all densities of N_1 due to the interaction with N_2, which leads to an increase in the intrinsic rate of increase *and* carrying capacity at the equilibrium density.

Density dependent effects of mutualism at high densities of N_2. This situation is represented in Fig. 3.7b–d. The following scenarios are conceivable. Mutualism acts symmetrically to increase per capita birth and decrease per capita death rates of the other partner (Fig. 3.7b).

$$b = b_0 - aN_1/(1 + mN_2), \tag{3.17}$$

$$d = d_0 + cN_1/(1 + mN_2). \tag{3.18}$$

This results in an increase in the carrying capacity for N_1 but does not affect its fitness (r). The more N_2 increases in abundance in this case the closer b and d are to their maximum and minimum values, b_0 and d_0 respectively, and the more will the unbounded equilibrium density be approached. This is of course an unrealistic scenario for any kind of mutualism. If only the birth rate is affected in the above way then a relationship like the one shown in Fig. 3.7c is achieved. Positive effects of mutualist N_2 on the birth rate of N_1 will increase both K and r of N_1. Interestingly, in the density independent relationship, mutualists with low growth rates, which can be achieved either by increasing death rates or by decreasing birth rates, benefit more than mutualists with high growth rates. However, this does not need to be the case in density-dependent

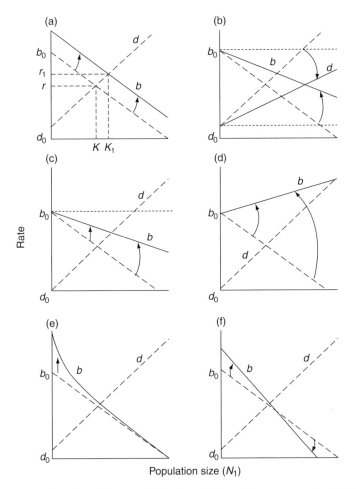

Fig. 3.7. Density dependent and density independent per capita birth and death rates in the logistic model in the absence of a mutualistic partner (= null model, dashed lines) and when influenced by mutualistic interactions (solid lines). (a) Density independent effects on mutualist; (b–d) high density effects; (e, f) low density effects of mutualist N_2. The population reaches a stable equilibrium at the intersection of the curves ($N = K$), where birth and death rates are equal. (After Wolin and Lawlor 1984.)

mutualistic relationships. Another high-density effect of mutualistic interactions is shown in Fig. 3.7d. Here, the per capita increase in birth rate of mutualist 1 actually exceeds b_0, e.g. if $mN_2 > a$, while death rates are as above:

$$b = b_0 - (a - mN_2)N_1, \tag{3.19}$$

$$d = d_0 + cN_1. \tag{3.20}$$

As a consequence, r may actually exceed r_0 with the carrying capacity also potentially increasing, but depending on the death rate. Those cases where mutualistic interactions are most pronounced at high densities of the interacting partners are probably not very realistic because such high densities might never be achievable due to constraints that operate independently of the mutualistic interactions. Similarly, the number of mutualists might be too small to provide benefits for the whole population. Therefore, Wolin and Lawlor (1984) also explored the effects of mutualism at low densities of N_2.

Density dependent effects of mutualism at low densities of N_2. If the per capita benefit provided by a given number of mutualists decreases exponentially with the density of N_1 such a relationship could be expressed as:

$$b = b_0 - aN_1 + mN_2e^{-\alpha N_1}, \tag{3.21}$$

$$d = d_0 + cN_1. \tag{3.22}$$

Fig. 3.7e shows that in this scenario the equilibrium density may increase but not necessarily the intrinsic rates of increase. The benefit is highest at low densities of N_1 and may actually exceed r_{max} but remains indifferent at higher densities of mutualist 1. Depending on the size of the constant α, the equilibrium density may be identical to K (large α) or exceed K (small α). Yet another way to describe low-density effects is to use linear density dependent effects such as:

$$b = (b_0 + mN_2) - (a + uN_2)N_1, \tag{3.23}$$

$$d = d_0 + cN_1. \tag{3.24}$$

While enhancing mutualist 1's birth rates at low densities this model predicts that at densities above K the birth rate will be lower than it would be in the absence of mutualist 2 (Fig. 3.7f). This is equivalent to saying that at low densities the partner acts as a mutualist, but at high densities it acts as a competitor or predator, which brings us back to the density dependent mutualism–antagonism continuum described by the previous models. When m approaches zero, this equation reduces to the high-density effect model considered in Fig. 3.7d, and if the constant u approaches zero the density independent model (Fig 3.7a) is retrieved. As in the previous models not only the birth rates but also the death rates can be affected (not considered here).

In spite of the fact that many of these interaction terms are constants or arbitrary functions they reflect the heterogeneity of possible mutualistic interactions and frame these interactions as acting directly on per capita birth and death rates. In this way a more mechanistic understanding of the dynamics of the interaction continuum is achievable. The interesting question, of course, is how close these different classes of density dependent and independent models are to the real world. Undoubtedly, density dependent effects are important in plant–insect mutualisms, where plant density can have positive or negative effects on pollination success (Beattie 1976). For homopteran–ant relationships there are a number of studies suggesting density dependent mutualism (Addicott 1979, Breton and Addicott 1992a, Sakata 1995, Morales 2000a). For example, in a study on the aphid *Aphis varians* feeding on *Epilobium angustifolium* Addicott (1979) showed that the effect of ant tending was lower at high aphid densities. Small, unattended colonies were more likely to decrease than attended colonies but large colonies showed no ant-attendance-dependent change in size. In a follow-up experiment Breton and Addicott (1992a) showed that the per capita growth rates of *A. varians* attended by *Formica cinerea* were significantly higher in low- than high-density populations (significant ant × density interactions). Observations on the predatory activity of *Lasius niger* on aphids also suggests that the density of a mutualist is a critical factor for determining whether *L. niger* either attends or eats aphids (Sakata 1995). This is reasonable behaviour if ants attempt to monopolize a honeydew resource and adjust the output to their requirements.

A shared feature of the models described above is that life-history traits, such as survival and fecundity, may vary with the density of a mutualist, but not explicitly over time. Thus, the predicted dynamics of these model populations are dependent only on the initial parameter values of these life-history traits. This conforms to the many life-history studies that have shown that fecundity and survival as well as the trade-offs between them vary with density. For example, Breton and Addicott (1992a) showed that with increasing aphid density per capita growth rates of *A. varians* significantly declined. Similarly, benefits of ant mutualisms clearly change with growing membracid or lycaenid densities over time, with smaller aggregations often benefiting more than large aggregations (Pierce *et al.* 1987, Morales 2000a). This might be because of intraspecific competition at high densities. However, fitness may also change over time independent of density, for example with varying environmental or seasonal conditions (Stadler 1995, 2004). Therefore, a more general approach is to include the effect of past environmental conditions and past trade-offs on the current and future fitness of organisms. This can be achieved by different methods. For example, age- and stage-structured models incorporate the idea that populations are made up of individuals belonging to different age or stage classes, which have different fecundity and survival rates. These models use Leslie matrices to describe how the number of adults at time t is a function of the number of eggs/juveniles at some time $t-x$ with all the different age and/or stage classes subject to particular life-history trade-offs.

Another interesting approach is incorporated in models including delayed life-history effects (DLHE models) (Beckerman *et al.* 2002). Delayed effects of environments on life histories are commonly reported in the life-history literature. Often these effects operate via the mother (parents) on the fitness of her offspring (Mousseau and Dingle 1991, Mousseau and Fox 1998, Hunter 2002, Zink 2003). Whether natural selection shapes maternal effects depends on the extent to which maternal environments and behaviour influence offspring phenotype and fitness. DLHE models include the idea that past environment affects fitness of future offspring by affecting the physiological condition of the mother, and that density might be one aspect but not necessarily the only one. For example, in many insects photoperiod, temperature, host availability and quality experienced by an ovipositing female will determine the probability of diapause, offspring size or offspring number. Wing polymorphism in aphids is often a direct consequence of the conditions experienced by an aphid such as overcrowding, plant growth stage (Lees 1967, Watt and Dixon 1981) or contact with natural enemies (Dixon 1958, Weisser *et al.* 1999). In *Drepanosiphum platanoidis* the reproductive rate in early autumn is not correlated with the number of aphids present at this time but dependent on the number of aphids present before they entered summer diapause. Crowding of aphids in spring and summer negatively affects their growth rates in autumn (Dixon 1975). Similarly, females of membracids exhibiting maternal brood care trade off current reproduction with future reproduction and exploit the protection service of ants. Their egg-guarding behaviour helps to establish ant attendance by increasing the ant to nymph ratio (Section 6.5; Olmstead and Wood 1990, Billick *et al.* 2001). After ant attendance is initiated and offspring are attended by ants, females abandon their first clutch and lay a second if environmental conditions permit. In the simplest terms maternal effects occur, when the phenotype or environment experienced by the mother affects the fitness of her offspring via some mechanisms other than the transmission of genes. Modelling delayed density dependence including maternal effects can be achieved relatively easily. The approach suggested by Beckerman *et al.* (2002) starts with a discrete maternal effect model for a single species.

$$N_{t+1} = N_t R \frac{x_t}{1 + x_t}$$
$$x_{t+1} = x_t \frac{M}{1 + N_{t+1}}$$
(3.25)

where N_x is the population size at a specific time t or $t + 1$ respectively, R is the rate of increase, which is sensitive to some measure of individual quality, x, or behaviour that is itself a function of population density. M is a species-specific

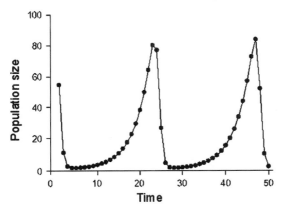

Fig. 3.8. Population cycles of a species experiencing delayed life-history effects. (After Beckerman *et al.* 2002.)

constant. The behaviour of this model is depicted in Fig. 3.8, showing the gradual increase in population size over time and sudden crash in numbers typical of many herbivorous insects such as aphids, coccids or membracids.

Using this model for two-species mutualistic interactions in which one partner is dependent on the other (e.g. an obligate myrmecophile) gives the following relationships:

$$N_{1(t+1)} = N_{1(t)} R_1 \frac{x_{1(t)}}{1 + x_{1(t)}},$$

$$x_{1(t+1)} = x_{1(t)} \frac{M_1}{1 + N_{1(t+1)}} N_{2(t)},$$

$$(3.26)$$

$$N_{2(t+1)} = N_{2(t)} R_2 \frac{x_{2(t)}}{1 + x_{2(t)}},$$

$$x_{2(t+1)} = x_{2(t)} \frac{M_2}{1 + N_{2(t+1)}}.$$

$$(3.27)$$

Thus, the population size of N_1 is influenced by delayed density effects, maternal effects and by the density of N_2. If no mutualist 2 is available ($N_{2(0)} = 0$) mutualist 1 cannot survive. If mutualist 1 and mutualist 2 are available, then they interact in a way shown in Fig. 3.9. These simple equations generate a surprising diversity of population fluctuations depending on the relative magnitude of the growth rates. For example, if mutualistic interactions cause an increase in R then the populations tend to fluctuate more widely with repeated crashes, typical of many aphid species. Like models incorporating delayed density dependence DLHE models predict increased fluctuations in population size, especially with increasing growth rates (a test of this

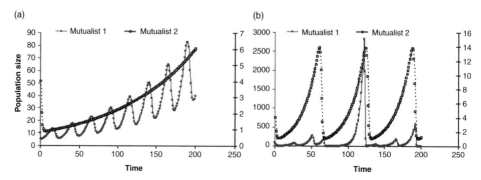

Fig. 3.9. Population cycles of two species experiencing delayed life-history effects. (a) $N_{1(0)} = 10$, $N_{2(0)} = 4$, $R_1 = 1.1$, $M_1 = 10$, $R_2 = 1.01$, $M_2 = 5$; (b) $N_{1(0)} = 87$, $N_{2(0)} = 4$, $R_1 = 1.2$, $M_1 = 10$, $R_2 = 1.05$, $M_2 = 5$.

prediction is provided in Section 5.2). For example, at $t = 122$ (Fig. 3.9b) there is a strong increase in the population size of both mutualists, suggesting a potential outbreak situation, at least of the more abundant mutualist. The reasons for this are twofold: (a) increase in R because of positive interactions and (b) delayed pay-offs in the benefits due to mutualism (Fig. 3.9b). If the delay is longer than one time unit then both populations tend to fluctuate even more.

This model provides a better fit to time series than delayed logistic models and it is possible to incorporate effects such as ant attendance, changing plant quality and seasonal changes in population size for different aphid species living on tansy (*Tanacetum vulgare*) (Stadler 2004). In a world that is dominated by variable environments at scales of individual plants, to patches and landscapes, models should include mechanisms describing how animals respond to these changes and their effect on future generations.

It is important to note that contrary to the Lotka–Volterra type models or the consumer–resource relationships suggested in biological market models, life-history (maternal effects) models do not assume that trophic (food web) interactions are the primary reason for density dependent fluctuations in natural populations. Rather they assume that the population level effects are a function of immediate effects on fecundity or survival, or delayed response to either density or life-history traits. This is important when evaluating hypotheses on the relative importance of predator–prey, competitive or mutualistic interactions versus life-history responses for population dynamics and community structure. However, it is unlikely that it will be easy to distinguish between strong trophic versus life-history effects incorporating trade-off and delayed density dependence, because competition and mutualism (as we shall show) will probably affect the same life-history traits of the partners, and mechanisms of delayed density dependence are particularly difficult to pin-point and are thus actively debated (Hunter and Price 1998, 2000, Turchin and Berryman 2000, Berryman *et al.* 2002). However, recognizing that mutualistic interactions can operate as regulatory mechanisms affecting population dynamics via delayed maternal effects would be a step forward. Many examples corroborating this statement will be presented in later sections.

IN SUMMARY, life-history models incorporate much information on the reproductive biology of individuals and show how it affects population growth. However, this approach has so far not been applied systematically to mutualistic interactions. That is, it remains to be shown how relative changes in growth rates or delayed density effects of both partners affect their population dynamics and how this ultimately affects the evolution of mutualistic interactions.

3.3.3 *Metapopulation models*

The members of a population are usually not distributed continuously in space, but are often highly aggregated as a result of variation in geophysical or ecological characteristics of the landscape. The aggregation of individuals within a particular area constitutes the local population and it is this unit of the larger regional population where most behavioural, genetic and ecological interactions occur. These local populations are assumed to be spatially separated from one another by unsuitable habitats, which do not sustain growth of individuals of the focal species. Almost all species have evolved mechanisms that enable individuals to cross unsuitable habitats at some stage in their life cycle, and thus most local populations are potentially connected to other populations through dispersal and migration. A central issue in the analysis of metapopulations is the frequency of migration, or connectivity among local populations. Many studies have shown that even a very limited amount of migration can have a profound effect upon the recipient population, with one or two successful migrants per generation causing an otherwise isolated population to behave as if mating between them is panmictic (Wright 1978, Stacy *et al.* 1997).

The concepts used in the previous sections were based on a perspective of closed populations that reach equilibrium sizes (zero growth isoclines). That is, interactions between members of a population as well as population regulation and density dependence occur within a defined spatial configuration (local population). In metapopulations the population is open and immigration and emigration between suitable patches occurs at demographically significant rates; that is, migration has a significant effect on local population size. This has

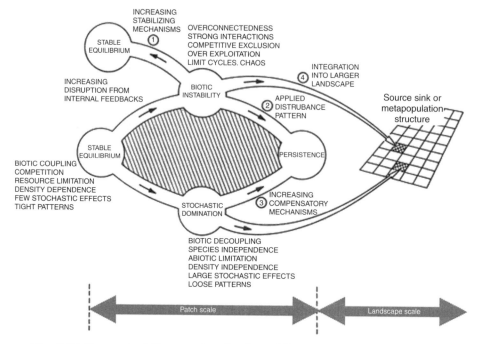

INCREASING
STABILIZING
MECHANISMS   OVERCONNECTEDNESS
            ① STRONG INTERACTIONS
STABLE   COMPETITIVE EXCLUSION
EQUILIBRIUM   OVER EXPLOITATION
                LIMIT CYCLES. CHAOS
                                            INTEGRATION
                                         ④ INTO LARGER
INCREASING   BIOTIC   LANDSCAPE
DISRUPTION FROM   INSTABILITY
INTERNAL FEEDBACKS   APPLIED
                                  ② DISTRUBANCE   Source sink or
                                    PATTERN   metapopulation
                                                       structure
STABLE
EQUILIBRIUM   PERSISTENCE
BIOTIC COUPLING
COMPETITION
RESOURCE LIMITATION
DENSITY DEPENDENCE
FEW STOCHASTIC EFFECTS   INCREASING
TIGHT PATTERNS   STOCHASTIC   ③ COMPENSATORY
                          DOMINATION   MECHANISMS

BIOTIC DECOUPLING
SPECIES INDEPENDENCE
ABIOTIC LIMITATION
DENSITY INDEPENDENCE
LARGE STOCHASTIC EFFECTS
LOOSE PATTERNS

Patch scale   Landscape scale

Fig. 3.10. Conceptual diagram showing four different hypotheses proposed to explain stability in ecological systems in the face of biotic instability and environmental stochasticity. Extrapolating these local processes to a larger landscape level reduces the importance of internal biotic feedback interactions and stochastic processes. (Modified after DeAngelis and Waterhouse 1987.)

important consequences as fluctuations in population size might become less dependent on local growth rates and density dependent processes than on the ability of organisms to travel across unsuitable terrain (Murdoch 1994). This shift in perspective raises the questions whether metapopulations can persist without density dependent feedback mechanisms (regulation) and how and at what spatial scale metapopulations are regulated. That is, it challenges the notion of density dependent regulation of populations (Sections 3.2.1 and 6.1).

In an early but still useful attempt DeAngelis and Waterhouse (1987) tried to visually organize equilibrium and non-equilibrium concepts in ecological models (Fig. 3.10).

For example, theoretical ecologists have examined a wide variety of local destabilizing and stabilizing mechanisms and have shown that stability in populations may be a consequence of functional relationships involving (Fig. 3.10): (1) searching time, predator interference or prey refuges; (2) disturbance that repeatedly interrupts positive or negative feedback effects; (3) compensatory mechanisms such as density dependent responses; (4) spatial aspects, involving source–sink or metapopulation features of populations. Interestingly, many models suggest that destabilization tends to increase with the strength of the interspecific biotic interactions. The Thompson (1924), Lotka (1925), Volterra (1926) and Nicholson and Bailey (1935) models are good examples. In a search for stability more complexity was added to these models. Eventually, this led to the identification of a number of stabilizing features, such as logistic prey growth, physical refuges for the prey, density dependent predator death rates (Murdoch *et al.* 2003) and also led to the recognition of potentially destabilizing factors like Type II, Type III functional responses, number of species, connectivity (May 1972, 1973) or time lags in density dependent processes, to name a few. These models seem to reveal potential biotic feedback instabilities inherent in complex natural species assemblages but simultaneously ignore other factors that counter the destabilizing influence of strong direct feedbacks. In many interacting populations, such a factor may be mutualism. In particular, the drawback of many of these models is the lack of an awareness that interactions may range from negative to positive at different times and at different places, with changing densities of the interacting partners. Such awareness, however, might be essential for understanding the outcome and dynamics of species interactions.

Stochastic events, which occur within a population via reproduction and mortality, and environmental perturbations also decrease stability (Fig. 3.10). While both biotic feedback and stochastic events might be disruptive to the system, increasing instability, the effects of both can compensate each other by a lack of a correlation between local populations in space. Certain types of stochastic events can moderate tightly coupled and unstable biotic feedbacks, and certain types of biotic forces can moderate stochastic impacts.

The idea of adding spatial scale was certainly a major step forward because it introduced a new perspective different from equilibrium thinking, and the explicit separation of within-patch and among-patch dynamics. The problem of scale is seen as central when attempting to unify population biology and ecosystem science (Levin 1992). The problems introduced by biotic and stochastic processes might be less severe if one accepts that real populations generally occupy large areas, so that the dynamics of any particular small unit may not be too important for the persistence of a species as a whole. Within a food web all components are subject to a spatial dimension, which means that a local catastrophic event (instability) and extinction of a local population is a negligible event (not affecting overall stability) at the level of the regional population. As concluded in previous sections, this does not mean that models that do not explicitly recognize the spatial scale are worthless, as many equilibrium models provide valuable insights into potential mechanisms

structuring populations and communities including those that contain mutualists. In addition, it is debatable which species are likely to have a metapopulation structure. In the case of ants and their insect partners it is unlikely that
the scale at which a metapopulation structure is found (if at all) differs substantially from species to species or is affected by their dispersal ability and
genetic structure. For example, for clonal organisms, or organisms with a high
degree of relatedness, such as ants, a patch, from the point of view of an
individual, might be substantially larger than the immediate host plant or
territory it currently feeds on or occupies. Thus, identifying selection processes in the continuum of antagonistic to mutualistic relationships in the
context of a food web structure requires thinking in terms of many interacting populations, each with its own scale of (meta) population structure. More
specifically, a metapopulation of ants is likely to be embedded in a metapopulation of aphids because of their greater dispersal capabilities and clonal
architecture relative to ants.

A good starting point is to introduce the competition-colonization trade-off
in order to understand how two species of closely related phytophagous insects
can co-occur on the same host plant when one is a myrmecophile and the other
is unattended. The simplest and most influential example of a spatial trade-off
model of species coexistence is that of Levins and Culver (1971). They showed
that two species can coexist in a metapopulation if one species is a superior
competitor and the other a superior colonizer. This model assumes that species
generally show a trade-off between competitiveness and dispersal ability. It is
summarized in two equations:

$$\frac{dp_1}{dt} = c_1 p_1 (1 - p_1) - m_1 p_1, \tag{3.28}$$

$$\frac{dp_2}{dt} = c_2 p_2 (1 - p_1 - p_2) - m_2 p_2 - c_1 p_1 p_2. \tag{3.29}$$

The term p_i represents the fraction of patches occupied by species i; c_i is the per
capita colonization rate, and m_i the per capita density independent extinction
rate of species i. Species 1 is considered to be the superior competitor because it
can colonize any habitat, which is not occupied by species 2 $(1 - p_1)$. Species 2,
in contrast, can only colonize empty sites $(1 - p_1 - p_2)$ and, as a result, is
displaced by species 1 $(-c_1 p_1 p_2)$. If we define a site as a single shoot of a
host plant this means that a myrmecophilous partner of ants could displace a
non-myrmecophilous partner, because with ants as allies it might be easier to
outcompete an unattended species locally. In spite of this local displacement
species 2 can still persist and coexist, i.e. on another shoot/branch of the same

host plant, if it has a higher rate of colonization ($c_2 > c^2_2/m_1$) provided that m_1 and m_2 are equal. This model suggests that ant-attended species are better competitors and non-myrmecophilous species better colonizers. This prediction is supported by a number of aphid species for which different degrees in the strength of associations are proposed (Fischer *et al.* 2001). For aphids feeding on the same plant but showing different degrees of ant attendance, the less myrmecophilous ones are removed by ants. Even though the mortality rates of these species need not be equal, the species that is removed by the ants needs to find shoots or hosts where the ant partner of the competing species is not present. In a comparative analysis of more than 100 aphid and lycaenid species the mobility (dispersal capability) of aphids is indeed negatively correlated with ant attendance, while for lycaenid larvae this trait is less important (Stadler *et al.* 2003) suggesting that the competition–colonization trade-off might at least partially explain the co-occurrence of often closely related species, with one being closely associated with ants and the other unattended. As will be shown in Chapter 5 these trade-offs are likely to vary in intensity over the season and need to take into account the costs of being myrmecophilous as well as the mortality risk in different habitats. The above model incorporating spatial aspects in niche partitioning has been modified in many ways, e.g. by incorporating dispersal–fecundity trade-offs (Yu *et al.* 2001), variation in patch density (Yu and Wilson 2001) or maintenance of biological diversity (Tilman 1994).

Nee and May (1992) extended the competition–colonization trade-off to analyse the effects of patch removal on the dynamics of metapopulations and regional abundance of two coexisting species. Again, there are two species, a superior competitor (by convention species 1) and inferior competitor, species 2, which is unable to invade a patch occupied by species 1. However, species 2 is the better colonizer of empty patches. Their model equations are:

$$\frac{dx}{dt} = -c_1xy + e_1y - c_2xz + e_2z, \tag{3.30}$$

$$\frac{dy}{dt} = c_1xy - e_1y + c_1zy, \tag{3.31}$$

$$\frac{dz}{dt} = c_2zx - e_2z - c_1zy. \tag{3.32}$$

The term c_i gives the colonization rates and e_i the extinction rates of species 1 and 2, respectively; x, y and z denote the proportion of empty patches, patches occupied by species 1 only, and patches occupied by species 2 only. This gives

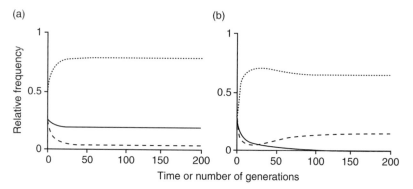

Fig. 3.11. Proportion of patches occupied by two species in a metapopulation setting. Species 1 (solid line) is the inferior colonizer but superior competitor; species 2 (dashed line) is the superior colonizer but the inferior competitor. Dotted line gives the relative proportion of empty patches. (a) $c_1 = 0.7$; $c_2 = 0.55$; $e_1 = e_2 = 0.45$. (b) $c_1 = 0.7$; $c_2 = 0.55$; $e_1 = e_2 = 0.45$ but with the number of habitable patches reduced by 20%. (After Nee and May 1992.)

the number of habitable patches h ($h = x + y + z$). Given that x, y, and z are greater than zero, the equilibrium solutions are:

$$x^* = \frac{1}{c_2}(hc_1 - e_1 + e_2),\qquad(3.33)$$

$$y^* = h - \frac{e_1}{c_1},\qquad(3.34)$$

$$z^* = \frac{e_1(c_1 + c_2)}{c_1 c_2} - \frac{e_2}{c_2} - \frac{hc_1}{c_2}.\qquad(3.35)$$

The necessary condition for the inferior competitor (species 2) to persist is if its colonization rate is larger than the colonization rate of species 1, given that the mortality rates are equal (Fig. 3.11a). More generally if:

$$c_2 > \frac{c_1 e_2}{e_1}.\qquad(3.36)$$

The inferior competitor might even persist and co-occur regionally, if it has a lower colonization rate provided it also has a lower extinction rate compared with species 1. Now, the interesting case is: What happens if the fraction of habitable patches is reduced?

If the proportion of available patches decreases, there is an increase in the number of patches occupied by the inferior species and decrease in the number occupied by the superior species (Fig. 3.11b). The reason is that the inferior species (fugitive species) is much better at dispersing and finding new sites, whereas the competitively superior species does worse when the proportion of available sites decreases. This simple scenario might be even more important for myrmecophilous phytophagous insects because their world is more patchily structured than the world of non-myrmecophiles as they need to find not only suitable host plants but also sites with ants or even an appropriate species of ant. Thus, given that habitats undergo successional changes the fraction of inhabitable patches for myrmecophiles might be more variable and subject to quicker decline with succession than for co-occurring non-myrmecophilous species, even when feeding on the same host plant. This is because ants also respond to changes in vegetation structure and might be less available in cool, temperate, tree-dominated sites.

Such competition-colonization trade-offs are not only relevant for phytophagous insects but are equally applicable to different ant species co-occurring in the same habitat. For example, tropical ant species inhabiting *Acacia* ('acacia ants') compete for the possession of host trees but still coexist at fine spatial scales (Yu *et al.* 2001, Stanton *et al.* 2002). Factors like heterogeneity in host-plant density, host availability (*h*) or stages of colony development are thought to contribute to the persistence of these ant-ant-plant mutualisms even in the presence of parasites. The population dynamics of aphids on fireweed is also thought to follow a metapopulation setting with some species showing high dispersal abilities while others do not (Addicott 1978b). For example, four species of aphids (*Macrosiphum valeriana*, *Aphis varians*, *A. helianthi* and *A. salicariae*) co-occur on shoots of *Epilobium angustifolium*. The three species of *Aphis* are attended by a variety of ants, while *M. valeriana* is not attended. Following the local dynamics of these aphids on more than 3800 shoots of fireweed revealed the species-specific abundance patterns and spatial colonization and extinction rates at 21 sites in Colorado, USA. From mid June to the end of August *A. varians* and *A. salicariae* reached higher numbers per shoot relative to *A. helianthi* and *M. valerianae* (Fig. 3.12). All species grew exponentially during the early period of population build-up with *A. varians* appearing early in the season and showing the highest initial growth rates.

The relative frequency of shoots occupied by these species also increased most quickly for *A. varians* and *M. valerianae*, with the latter showing the largest number of shoots occupied at the end of August (Fig. 3.13). The patterns for *A. helianthi* and *A. salicariae* were similar, showing a monotonous increase in the proportion of occupied shoots; however, they appeared much later than the other two species. This suggests that new colonies were initiated throughout summer and especially by the unattended, highly

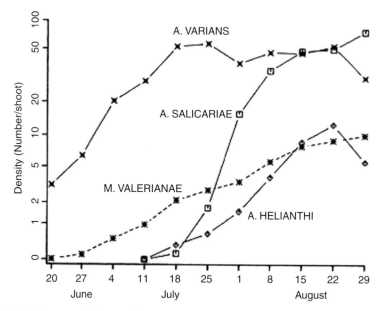

Fig. 3.12. Seasonal changes in population densities of four species of aphids on fireweed. Density is calculated as the total number of individuals of each species divided by the total number of shoots. (After Addicott 1978b.)

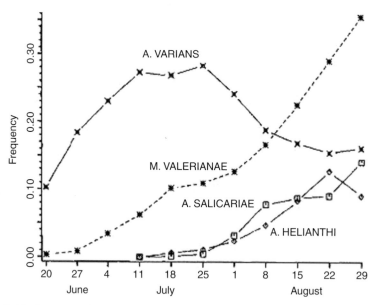

Fig. 3.13. Seasonal changes in the frequencies of shoots of fireweed occupied by four species of aphids. Frequency is calculated as the total number of shoots occupied by a given species divided by the total number of shoots each week. (After Addicott 1978b.)

mobile species, *M. valerianae*, which seeks out new shoots, especially when predator pressure is most severe.

To follow the dynamics of colonies on individual shoots Addicott (1978b) compared the size of a population in one week (t) with that in the next week ($t+1$). Percentiles, representing the cumulative probability of a population being less than some particular density at time $t+1$ were calculated (Fig. 3.14). This way of viewing changes in population dynamics reveals a number of things. (1) Even medium and large populations went extinct and half of the small colonies went extinct from one week to the next. For example, of the colonies of *A. varians* with only one individual per shoot only 50% were present the next week. Of those with between 12 and 20 individuals, approximately 25% went extinct. These figures show high rates of extinction at all density levels and a high degree of turnover of individual populations. (2) The performance of the local populations at any given density level frequently fell below the equilibrium line (indicating no change between successive population surveys), implying that most populations declined or did not change at best. This means that continued growth of the regional populations could occur only if new populations were initiated. (3) Even for those colonies that performed very well (i.e. 95th and 99th percentiles) there were densities at which the colonies did not increase. These levels were highest for *A. salicariae* and *A. varians* and lowest for *M. valerianae*.

At the highest densities *M. valerianae* reached the equilibrium line with a positive slope regardless of the percentile chosen, while *A. salicariae* reached it with a negative slope, indicating a possible overshoot of the equilibrium density level. This might be a result of a closer association of *A. salicariae* with ants leading to a stronger increase in population size (see Fig. 3.12), visible damage to the host plant and subsequent crash in numbers below the equilibrium line. Consequently, this species showed the largest fluctuations in density. Most likely the colonies that did very well (95th percentile and above) may have gained several reproductive adults via immigration from nearby shoots. *Aphis salicariae* overexploited the host shoots followed by dispersal. As for other facultative myrmecophilous *Aphis* species different rates of ant attendance might result in different extinction rates, especially of small, slowly growing colonies. The highly mobile and unattended *M. valerianae*, in contrast, disperses from one shoot to another, particularly if the colony is disturbed by natural enemies or rain, thus occupying a habitat more evenly, but with smaller local colony sizes.

The important question to consider here is whether these aphid species form a metapopulation with an individual shoot of a fireweed ramet as the local patch occupied by an aphid clonal population. Addicott (1978b) concludes

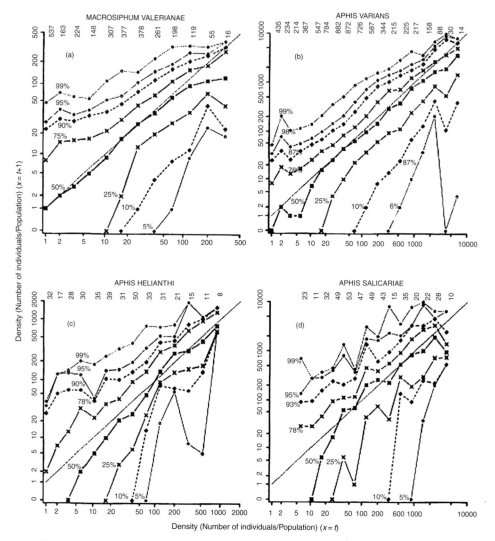

Fig. 3.14. The week-to-week changes $(t, t+1)$ in the abundance of four species of aphids on fireweed. Each line represents a different percentile of the distribution of the population size within different size classes in a given week. The number of colonies in each of these size classes is given at the top of each figure. The diagonal equilibrium line indicates where population size does not change from one week to the next. (After Addicott 1978b.)

that the aphids of a given species on individual shoots of fireweed constitute local populations. He argues that this is because some colonies can persist for several weeks on individual shoots of fireweed in spite of the fact that many are eliminated by natural enemies and that a local population is generally defined by the spatial unit and that the mobility of individuals is high within relative to

between units. 'If sites rather than shoots are the unit of a local aphid population on fireweed, then there should be a high degree of movement between
shoots, resulting in most shoots being occupied by at least a few individuals.'
Although this is a valid argument it defines a metapopulation in a static way
and largely ignores the biology of the associated species. The spatial structure
of a community of phytophagous insects should be a function of the differences among species, their genetic structure, the structure of their host plant
and their response to ants. For aphids this means that individuals with an
identical genetic structure form an organism, which occupies not only one
shoot of a plant, but potentially many shoots and even different individuals
of the host plant. Similarly, in the case of fireweed or tansy, a single shoot does
not constitute a single genetic individual because fireweed produces stolons
potentially interconnecting several hundreds of shoots of a single fireweed
plant. This makes a local patch much larger than the immediate feeding site
might suggest. Therefore, Weisser (2000) argues that for aphids on tansy
(*Tanacetum vulgare*) the genet rather than the ramet is the local patch. In
addition, being associated with ants imposes an additional layer of metapopulation heterogeneity or patchiness on the focal organisms. In the aphid–
ant–fireweed system, *M. valerianae*, which is unattended, was negatively
affected by ants, the facultative myrmecophiles *A. varians* and *A. helianthi*
were affected by ants in a density dependent way and *A. salicariae* was not
affected (Addicott 1979). Dispersal by alates is often restricted when attended
by ants (El-Ziady and Kennedy 1956, Johnson 1959), which affects the aggregation of aphids. This means that the gap between patches might actually
increase for attended aphids making it more difficult for them to find another
patch. According to the competition-colonization hypothesis this ant-induced
handicap can only be compensated for if the mortality risk declines.

Another source of evidence that moth and aphid populations might be less
prone to metapopulation dynamics comes from a study on the spatial synchrony in abundance of these taxa in the UK (Hanski and Woiwod 1993).
They tested the assumption that species with high maximum growth rates (e.g.
aphids) should show a more intrinsically generated variability than species
with low growth rates (e.g. moths). Using samples from British moth and
aphids collected at 57 (Rothamsted Insect Survey's network of light traps) and
21 (suction traps) localities throughout the UK they could not find a positive
relationship between maximum growth rates and the degree of spatial synchrony. For aphids this relationship was even negative. Spatial synchrony was
generally high and considerably higher in aphids than in moths and it declined
with increasing distance between conspecific populations. Therefore, it was
concluded that the observable large-scale patterns in spatial synchrony in

moths and aphids are most likely due to the spatially correlated variability in weather conditions and not due to migration. In particular, aphids showed a greater sensitivity than moths, probably due to their high growth rates being more sensitive to weather conditions.

There is a growing number of studies that incorporate the spatial dimension of mutualistic interactions, which suggests that the perceived habitat fragmentation of a species depends on its trophic position (Tscharntke and Brandl 2004). As a consequence, communities should be viewed as assemblages of species with different spatial strategies. It follows that phenomena occurring at spatial levels other than the local population must be considered in order to gain a good understanding of the evolutionary constraints and mechanisms influencing the dynamics in mutualistic systems. However, as was shown in the overview of models, assuming a closed structure, there is still much to learn from studies on the population dynamics of local populations. The challenge will be to find a balance between the incorporation of significant local processes and population data on large spatial and temporal scales for understanding the antagonistic and mutualistic trophic interactions across fragmented landscapes. Some interesting recent developments are described below.

Recent theoretical work suggests that populations subject to Allee effects have greater stability when random dispersal exists among local populations, because migrants can boost declining populations (Amarasekare 1998a, b, Greene and Stamps 2001, Greene 2003) in a metapopulation setting. Allee (1938) suggested that positive interactions may exist among individuals of a population at low densities facilitating growth and reproduction (increasing per capita growth rates), while at higher densities per capita growth rates decrease as a result of competition. Fitness thus is a function of population size and it might also imply that there exists an optimal population size and any deviation should lead either to aggregation or dispersal of individuals. However, the actual form of such a function might vary substantially depending on species, genetic structure and patch configuration. As a result, there is little empirical support for this seductive idea.

In addition to the Allee effect it is important to know how the dispersal of the local mutualists influences the local source communities. In a model similar to the basic competition-colonization model described above it was suggested that if dispersal involves surplus individuals, the loss of which does not affect the reproductive output of the source communities, then the persistence of sink communities is guaranteed as long as the survivorship of long-distance dispersers exceeds a certain threshold (Amarasekare 2004). Alternatively, if dispersal involves emigrants that currently contribute to the growth of the source community, persistence of the source communities requires that emigration does not exceed a certain upper threshold. In association with the Allee effect too many dispersers from the mutualist community can lead to large fluctuations in population sizes and landscape-wide extinction (Amarasekare 2004). That is, if the benefit to sink populations is outweighed by the cost of dispersal to source populations, then mutualistic communities that are linked by dispersal may experience a significant loss compared with those that are isolated. This scenario, however, would only hold for highly specialized mutualistic systems consisting of two species, one of which is relatively immobile and depends on a relatively mobile species. Originally described for plant-pollinator mutualistic systems such a scenario might be more relevant to ant–coccid systems compared with ant–aphid or ant–lycaenid systems. However, in insects there are often morphs that are specialized for dispersal and do not contribute to the reproductive output of the source population at all. This polymorphism is especially strong in coccids, membracids and aphids. We know of no case in ant–ant-partner obligate mutualisms where a too large emigration (large net loss) of individuals from the source population would eventually lead to a regional extinction of the entire mutualism. Often emigration is density dependent, with only a fraction of individuals dispersing. On host plants with genetically mixed aggregations of phytophagous insects the effects of dispersal on the dynamics of mutualistic interactions might not be as simple as described above.

Given that dispersal is risky it is likely that there is a conflict of interest between different genotypes. Staying on a host plant avoids dispersal risks but increases intra- and interspecific competition. Those genotypes that disperse might suffer a greater fitness reduction than those that stay. In other words, the dispersing genotype accepts the risk of dispersal, while the staying genotype benefits from reduced population size and reduced competition. It is unlikely that such a scenario would evolve. It is, however, more likely that an evolutionarily stable strategy will evolve, with dispersal in a fragmented landscape being determined by species-specific life-history attributes and density regulation within-patch population dynamics, leading to species-specific dispersal rates and all species showing some degree of dispersal (Parvinen *et al.* 2003). Therefore, it is unlikely that models incorporating Allee effects and dispersal (Greene 2003, Amarasekare 2004, Yamamura *et al.* 2004) are applicable to many insect mutualisms.

IN SUMMARY, mutualistic communities, like antagonistic/exploitative communities, contain groups of species, which potentially interact and are spatially segregated into distinct patches connected by dispersal. The trade-offs in species performances along multiple axes incorporate many of the traditionally perceived traits, such as density dependence, maternal effects, differences in resource exploitation or temporal variation in environmental conditions and the differential abilities to thrive under these varying conditions necessary for coexistence. Space effectively increases the dimensionality of a community, suggesting that interactions occur within local communities and differential dispersal abilities affect the persistence of local communities in a spatial setting. Hubbell (2001) has challenged the view that trade-offs are necessary to understand broader patterns of species diversity and relative abundance (see also Bell 2001). His neutral model assumes that drastic simplifications of the many processes that shape ecological communities can be made because life-history trade-offs should result in the same level of fitness for all groups of organisms exposed to the same environment. This theory was termed *neutral* because it essentially assumes that all individuals in a community are equal with respect to reproduction and probability of death (fitness equalization). It has achieved some success mainly when applied to diverse tropical forest communities. However, in contrast to the neutral theory, we believe that there is considerable evidence showing that trade-offs are meaningful relative to the environmental context and that different degrees of mutualistic associations are one of many environmental traits along which partners of ants can segregate (Kneitel and Chase 2004). It remains to be determined whether population regulation is achieved either by local stabilizing mechanisms (natural enemies or competition for resources) or via metapopulation dynamics and in which way mutualistic systems deviate or are similar to antagonistic systems.

4

Mutualisms between ants and their partners

4.1 Phylogeny and feeding ecology

Ants collect liquid food from caterpillars of the families Lycaenidae, Riodinide and Tortricidae (Maschwitz et al. 1986, Hölldobler and Wilson 1990, DeVries 1991a, Pierce et al. 2002), Sternorrhyncha (scale insects, aphids, white flies), Auchenorrhyncha (cicadas, leafhoppers, planthoppers) often summarized under the term 'Homoptera' (Buckley 1987, Wood 1993, Gullan and Kosztarab 1997, Delabie 2001, Stadler and Dixon 2005) and Heteroptera (Maschwitz et al. 1987, Hölldobler and Wilson 1990). In addition, they collect nectar from plants (Davidson et al. 2004, Oliveira and Freitas 2004). Many of these mutualisms are facultative and unspecialized, but the common denominator of all these associations is that they are driven by sugary excreta that are attractive to ants. Otherwise the life histories of these partners of ants are highly diverse. A crude characterization of the different taxa is that Sternorrhyncha and Auchenorrhyncha are often gregarious during some stage of their development, while many butterfly larvae have a more solitary lifestyle. Dispersal may occur in the early instars, as in many coccids (crawlers), while all other partners of ants disperse as adults. The beginning of the interactions between ants and other insects, in particular between ants and homopterans, dates back to the early Tertiary, because fossils in Baltic amber suggest that associations between aphids and *Iridomyrmex* spp. have existed since the early Oligocene (Wheeler 1910, Hölldobler and Wilson 1990). The ant species that evolved an ability to collect sugary excreta usually belong to just three subfamilies: *Dolichoderinae*, *Formicinae* and two genera of the *Myrmicinae* (Nixon 1951, Carroll and Janzen 1973).

There often is a tendency in the review literature to focus on these specialized associations (e.g. obligate myrmecophily) rather than ask what drives the large majority of interactions. This is understandable as it makes the subject more interesting, but at the same time is unfortunate because it prevents a deeper understanding of the magnitude, dynamics and frequency of different levels of associations dominating in nature. Another problem is that the relationships between ants and plants or other insects are often perceived to be based on trophic interactions with ants actively collecting sweet resources from their partners. This interaction is often termed 'trophobiosis' according to Wasmann (see e.g. Hölldobler and Wilson 1990) merging the notions of trophic relationships and symbiosis. However, trophobiosis is a catchword widely used in the ant literature, which tends to oversimplify and mask the complex associations between ants and many of their insect partners. In particular, this term implies an ant-centred point of view, which we believe is inadequate to describe the selection pressures the partners impose on each other. In addition, these associations are in most cases neither as spatially close as the term symbiosis might imply nor are interactions purely determined by trophic relationships. As we will show in the following sections, viewing mutualisms as another form of *mutual* exploitation requires an understanding of the range of adaptations evolved and fitness costs and benefits experienced by each partner well beyond the trophic aspects. As already suggested on theoretical grounds, and as we shall see below, associations between ants and their partners are influenced by physiological properties, local density dependence and larger scale processes implicit in local habitat and regional landscape characteristics. Reducing these relationships to a 'trophobiosis' does not do justice to the dynamic nature of the interactions. We therefore do not use this term when describing the dynamic relationships between ants and their insect partners nor do we use the term symbiosis to describe mutualistic interactions between free-living organisms.

Ants have existed since the upper Cretaceous, some 100 million years ago, and now comprise some 8000 species with the greatest diversity in the tropics (Hölldobler and Wilson 1990). They show division of labour and eusociality, which enables colonies to sustain high population densities for long periods of time (Wilson 1987, Hölldobler and Wilson 1990). Ants are amongst the best-studied insects and there is a good understanding of their feeding ecology, reproductive biology and evolutionary history. Most likely, there are two primary reasons for interacting with individuals of other species without killing them. One is to obtain resources, which are otherwise difficult to obtain, and the other is to gain nutrients/energy over a longer period of time. In all mutualisms involving insects there is always one partner that receives energy and/or nutrients from the other partner, while the latter receives some other service. Ants are certainly a good example of this. In spite of the fact that it is mostly ants that are the recipients of energy/nutrients, surprisingly much of the nutritional ecology of major ant phyla remains uncertain because observations are often restricted to short time periods or specialized ant genera like the fungus-growing *Attini* (Mueller *et al.* 2001), wood ants (*Formica* spp.), which are major predators in European forests (Gösswald 1989a, b), or fire ants (*Solenopsis* spp.), which are major urban or agricultural pests. Nevertheless, an impressive list of food types is available for different species of ants. For example, different ant species are known to feed on the sugary excreta (honeydew) of aphids, coccids, lycaenids, membracids and psyllids (Hölldobler and Wilson 1990), plant juices and nectar of trees, shrubs, fruit juices, seed and arthropod prey or carcasses (Levieux and Louis 1975, Levieux 1977, Retana *et al.* 1988, Azcarate *et al.* 2005, Davidson 2005, Fischer *et al.* 2005). In spite of these detailed observations there is a paucity of quantitative data showing the relative extent to which ants feed on one food type or another and in which way dietary requirements change over the course of time. For example, Stradling (1987) argues that ants are carnivorous but supplement their diet with other foods, suggesting that most species of ants are omnivorous. Carroll and Janzen (1973) claim most species of ants are scavengers rather than predators. Tobin (1993) disputes the view that ants are fundamentally carnivorous, a category in which he includes both predators and scavengers. He believes that the ants, by virtue of their abundance, must be 'the dominant primary consumers in most temperate and tropical ecosystems'. A further indication is that the large biomass they are able to achieve in tropical forests (Wilson 1990) suggests that they feed to a large extent on nectar, honeydew, food bodies and fungi rather than on insect prey (Tobin 1991, Davidson 1997). That is, they are more likely to be primary consumers than predators or scavengers. Initially this appears difficult to believe because like all adult aculeate Hymenoptera, adult ants have a pair of mandibles that appear utterly unsuitable for feeding on liquid foods. Yet, the availability of liquid food must have resulted in ants developing the ability to collect, transport and process energy-rich resources. Clearly, this affected their evolution (Wilson and Hölldobler 2005). Associated with an improved ability to collect nectar/honeydew/plant juices must have been organizational (behavioural) abilities associated with exploiting and monopolizing/defending these resources against conspecifics and other exudate feeders. An increasing success in collecting liquid food might have eventually brought the ants closer to the sources ultimately resulting in some form of by-product – protection against predators. Feeding patterns of ants are probably more diverse than previously thought and should be viewed in their ecological context; that is, they are highly conditional. A useful tool to elucidate resource partitioning in a complex community consisting of ants and partners of ants with multiple sources of honeydew/nectar and prey is the stable isotope technique. The isotopic C and N signature of whole organisms provides integrated information on the types of food consumed by larvae and thus some information on the trophic position of a particular species within a community. The analyses of ant communities in tropical rainforests of Australia provide convincing evidence that honeydew and nectar from extra-floral nectaries (EFNs) are highly important resources for arboreal ant communities (Blüthgen *et al.* 2003). In contrast, in ground-foraging ant species predation is more pronounced, as shown by the higher $\delta^{15}N$ isotopic signatures of their workers compared with those of arboreal species. Similar conclusions were derived from an analysis of ant communities in lowland tropical rainforests in Peru and Borneo, showing that arboreal ant communities are mostly 'herbivores' feeding on nectar secretions and insect honeydew (Davidson *et al.* 2003).

To summarize the feeding ecology of ants, tropical ant communities appear to be dependent to a large extent on carbon resources derived from plants and homopteran partners. Many of these species are less predacious than previously thought. This dependency on C-rich and N-poor food likely introduces two problems, one for the ants and one for the plants. If ants are to exploit a nutritionally imbalanced resource, such as the honeydew of homopterans or nectar secretions, they face osmoregulatory problems similar to those of homopterans. One way out of it might be to develop symbiotic relationships with gut microbes in order to exploit this C-rich resource. There is some evidence that this is the case (Boursaux-Eude and Gross 2000, Davidson *et al.* 2003). From the point of view of plants the ant species that are more primary than secondary consumers pose a particular problem, because it means that their defence system based on EFNs does not work or at least does not work effectively in an environment with many non-predatory ant species. Extrafloral nectaries may essentially become maladaptive in a honeydew-rich world exploited by honeydew specialists.

Morphological adaptations of the gut of ants for transporting large quantities of honeydew are essential for aphid–ant relationships. In the subfamilies Formicinae and Dolichoderinae the development of a proventriculus and flexible gasters enables many species to gather large quantities of fluid rich in carbohydrates from plants or homopterans (Eisner 1957, Davidson 1997, Davidson *et al.* 2004). This digestive organ is situated posterior to the crop and regulates the flow of food. The proventricular bulb pumps liquid from the crop into the midgut and 'prevents' the posterior flow. Only in the Dolichoderinae and Formicinae does the proventriculus passively occlude the passage of food through the gut, which allows the associated musculature to be reduced (Eisner 1957). In this way ants are able to control the movement of honeydew from the crop, the 'social stomach', into the midgut, where it is digested. In addition, in order to store large quantities of honeydew and carry it to the nest, ants need to be able to expand and contract their gaster according to the changes in the volume of honeydew collected (Kunkel *et al.* 1985, Taylor 1978).

Aphids evolved their greatest diversity in the temperate regions (Dixon *et al.* 1987) with about 4000 described species (Eastop 1973, Remaudière and Remaudière 1997). Feeding on plant sap is a very ancient form of herbivory dating back to the early Devonian some 400 million years ago (Labandeira 1997, van Ham *et al.* 2003, Wilkinson *et al.* 2003). This might be because plant sap is easier to digest than other plant material as it contains no fibre and relatively little by way of secondary compounds, which are stored in leaves and wood rather than in plant sap. However, they had to evolve methods of exploiting this resource, circumventing the nutrient imbalance and maintaining a high growth rate. One of the early developments was the formation of a symbiotic intracellular relationship with *Buchnera aphidicola* some 80–150 million years ago. This bacterium is thought to upgrade the non-essential amino-acids in plant sap to essential amino acids (Douglas 1989, Douglas and Prosser 1992, Douglas *et al.* 2001). Phloem sap usually contains high concentrations of sugar but little nitrogen (Kiss 1981, Kunkel *et al.* 1985), making aphids and other xylem/phloem sap feeders a highly nitrogen-limited group of insects. The high levels of sugar:amino acids and non-essential:essential amino acids are a problem, as are the overall high concentrations of low molecular weight sugars in the phloem sap. In order to reduce osmotic pressure and avoid dehydration aphids convert simple sugars to oligosaccharides in their gut until the contents are isosmotic with the haemolymph (Rhodes *et al.* 1997, Ashford *et al.* 2000, Douglas 2003). This is achieved in a through-flow system and is highly adaptable to the actual sugar concentrations in the phloem sap. For example, (Rhodes *et al.* 1997) showed that oligosaccharide synthesis occurs in the stomach of *Acyrthosiphon pisum* and the extent of the synthesis of high molecular weight sugars depends on the sucrose concentration in the sap. When the sucrose concentration is low honeydew contains mainly mono- and disaccharides, but when it is high oligosaccharides dominate. This suggests that the 'filter chamber', which is found in some aphid species, serves to dilute rather than concentrate the ingested sap and so further protect the aphid from dehydration (Goodchild 1966, Ponsen 1991). Many diverse forms of associations have evolved between ants and their partners belonging to several families of homopterans (for a general overview see Delabie 2001), with some of the associations highly specialized, but most are facultative or opportunistic.

Coccids (soft scale insects) occur in Tertiary amber and are assumed to have evolved during the early to mid Mesozoic (Gullan and Kosztarab 1997) and there are some 7500 species (Gullan and Martin 2003). Molecular analysis of the 18 S ribosomal DNA suggests that they are a sister group of the Aphidoidea (von Dohlen and Moran 1995) supporting earlier views of relationships between Auchenorrhyncha and Sternorrhyncha based on morphological traits (Hennig 1969). Two evolutionary phases might be distinguished. A primary phase from Permian to Jurassic during which coccids may have fed on plant roots, remains of plants and associated bacteria and a secondary radiation beginning in the Cretaceous when coccids became predominantly herbivores of above-ground plant parts (Gullan and Kosztarab 1997). This epigeic and hypogeic origin might explain several morphological features typical of coccids, such as wingless females, sessile way of life, legs that are more useful for digging than climbing and an antenna, which is a tactile organ with no chemoreceptors (Koteja 1985). A peculiarity of the life cycle of coccids is that females are neonate reaching reproductive maturity in the fourth instar, unlike males, which are winged and reach sexual maturity after the fifth moult. The marked sexual dimorphism, resulting from the neoteny of adult females and holometabolism of males, evolved only once and is shared by all extant scale insects (Gullan and Kosztarab 1997). Scale insects primarily feed on either the phloem or parenchyma cells and produce honeydew, which is discarded or collected by ants. Most coccids are monophagous but highly polyphagous species are also known (Gullan and Kosztarab 1997).

Like lycaenids and aphids, only a few taxa of mostly tropical and subtropical scale insects evolved obligate relationships with ants. Mostly these species are inquilines that can only survive in an ants' nest either within the chambers (domatia) of ant plants or in shelters built around the colonies (Flanders 1957). However, ants may also consume the coccids that live in their nests, especially if the ant colony is heavily dependent on the coccids for food (Carroll and Janzen 1973). Otherwise most associations of free-living species are facultative and unspecific (Way 1963, Gullan 1997).

The *treehopper* family Membracidae have their highest diversity in the New World tropics and comprise about 3100 species (McKamey 1998). They are distinguished from other homopterans by their often bizarre spike- or shield-like pronotum extending forward. Membracids are thought to have evolved in the Tertiary in the New World reaching the Old World through dispersal (Dietrich and Deitz 1993), now with just three species in Central Europe. This insect group has attracted considerable attention because of a rather unusual combination of features: subsociality, which is the maternal care of eggs and nymphs, mutualisms with ants (Wood 1993) and acoustic communication (Cocroft 1996, Cocroft and Rodriguez 2005). Most likely maternal care arose independently several times in various lineages and there are a few cases where maternal care has been lost. The evolution of maternal care in Membracinae has a strong phylogenetic component and thus is probably less evolutionarily labile than previously thought (Lin *et al.* 2004). Maternal care could also affect the associations with ants and some good examples are described in Section 4.2.4. Interestingly, in a survey of New World membracids Wood (1993) found that ant-attended membracids show fewer defensive behaviours compared with unattended species. This is similar to ant-attended aphids, which are often gregarious, less mobile and have shorter siphunculi (Mondor *et al.* 2002). One possible explanation for a reduction in defensive traits is that ant tending provides their partner with some enemy-free space. However, this poses a problem for partners of ants if they have to compete (at least temporarily) for the services of ants, because the lack of defensive structures may impose significant association costs.

Treehoppers are phytophagous insects with piercing and sucking mouthparts that they use to feed on the phloem and xylem of their hosts. In some species early nymphs are unable to penetrate the host plant and reach the phloem bundles. In these cases maternal care includes an unusual behaviour in which females use their ovipositor to create punctures in plant tissue and the resulting exudates are fed on by the early instars (McKamey and Deitz 1996). Membracids excrete surplus plant fluids as honeydew, which is either flicked off with their anal whip or taken up by attending ants. As with other ant-attended insects, the presence of ants greatly increases the survivorship of nymphs (Wood 1977). In some cases the treehoppers pass the parental care to ants in order to lay a second clutch of eggs (Zink 2003).

These associations evolved independently and mutualistic interactions with ants do not seem to be constrained to particular homopteran or heteropteran taxa as, although not analysed phylogenetically, the frequency of ant attendance does not vary strikingly between groups, probably suggesting multiple origins of myrmecophily (Bristow 1991, Gullan and Martin 2003, Stadler *et al.* 2003).

The Lycaenidae probably evolved in the mid Cretaceous about 100 million years ago (Eliot 1973) and are most diverse in tropical regions. They are distinguished in many ways from the partners of ants described above. They are holometabolous insects, exclusively sexually reproducing and do not feed on plant sap. Contrary to aphids or coccids, in lycaenids the frequency of ant attendance appears to vary within different groups, because this taxon shows a clear geographic pattern in the levels of association with ants. For example, obligate associations are far more frequent in the southern than in the northern hemisphere (Pierce 1987, Fiedler 1997a, Eastwood and Fraser 1999) (Table 4.1).

According to this data obligate associations are prevalent in Australia (39%) and South Africa (59%), while in the Nearctic only 2% of the lycaenid

Mutualisms between ants and their partners

Table 4.1. *Geographic distribution of the major lycaenid taxonomic groups (subfamilies and tribes) in different regions and prevalence of associations with ants*

Taxonomic group	Zoogeographical region					
	Australian	Afrotropical	Oriental	Palaearctic	Nearctic	Neotropical
Poritiinae						
Poritiini			+++			
Pentilini		+++				
Liptenini		+++				
Miletinae						
Liphyrini	+	++	+			
Miletini	+	+	+++			
Spalgini		+	+	+	+	
Lachnocnemini		+++				
Curetinae	+			++	+	
Lycaeninae						
Theclini	++++	+	++++	+++	+++	++++
Aphnaeini		++++	++	++		
Lycaenini	+	+	++	+++	++	+
Polyommatini	+++	++++	++++	++++	+++	+++
Riodininae	++	++	++	++	++	++++

+, low representation in the region (usually 1–4 species); ++, moderate representation (5–30 species); +++ high representation (31–100 species); very high representation (>100 species).

Source: After Pierce *et al.* (2002). Data were compiled from Eliot (1973), Fiedler (1991) and DeVries (1997).

species are considered obligate myrmecophiles and more than 80% of the species are unattended (Pierce *et al.* 2002). The causes of this pattern appear to be (1) the systematic composition of the major taxonomic groups in these regions and (2) differences in environmental factors. For example, the high degree of obligate associations in the southern hemisphere is largely determined by the dominance of the *Theclini* and *Aphnaeini* in these regions and the low preponderance of non-ant-associated subtribes of *Theclini*. In the Nearctic and Palaearctic non-attended *Polyommatini* dominate the lycaenid faunas (Pierce 1987, Fiedler 1991). Environmental factors are less easy to identify as determinants of levels of associations with ants because detailed phylogenetic information is needed to disentangle faunal composition and environmental cues. For example, host plant relationships of lycaenid butterflies do not appear to be more specialized in the tropics than temperate regions, while the association with ants seems to be more specific in the tropics (Fiedler

1997a). Similarly, there is equivocal support for the idea that because of the nutritional requirements of myrmecophiles, ant-attended lycaenids prefer to feed on nutrient-rich plants, such as legumes (Pierce 1985, Fiedler 1995). This 'plant permissive hypothesis' (Bristow 1991) is also suggested to be important in other associations between ants and myrmecophiles, but, as we shall show below, generally has little empirical and theoretical support.

Lycaenids have developed a number of organs to facilitate communication with and manipulation of ants. These organs, like the pore cupola organs (PCOs), tentacle organs (TOs) or dorsal nectary organs (DNOs) are found on most lycaenid larvae associated with ants. They function to appease ants or attract attention by secreting volatiles when the larvae are disturbed. Nectary organs produce sugary secretions and are of primary importance in ant–lycaenid communication and for rewarding attending ants (Malicky 1969, Leimar and Axen 1993, Axen *et al.* 1996). These 'honeydew' organs are exocrine glands, which produce honeydew in a way very different from homoptera. Functionally, these organs are closer to the extrafloral nectaries of plants than the honeydew excretory system of aphids. Secondary loss of these organs is often associated with a reduction in ant associations, indicating that it is costly for them to secrete nectar.

4.2 Associations of ants with nectar/honeydew producing partners

4.2.1 Ants

There are a number of ecological features that ants share with aphids. For example, ants and aphids inhabit many different habitats and are abundant (Wilson 1987, 1990). They are colonial, female-dominated societies with a modular structure, which means that the death of a single individual does not affect the survival and genetic structure of a colony. Like aphids, many ant species might be called food specialists, depending on a few types of food, such as carbohydrates, but omnivory – the living on a variable mixture of insect prey, aphid honeydew, or nectar from plants or butterflies – is thought to prevail, especially in temperate regions (Hölldobler and Wilson 1990). In addition, different members of a colony perform different tasks, such as migration, defence and reproduction. In other words, ants show a complex division of labour.

Unique features of ants are that their societies are long lived (relative to the lifespan of an individual ant) with distinct dominance hierarchies between different species. Their abundance is believed to be only limited by the availability of energy (Kaspari *et al.* 2000a). The energy limitation hypothesis suggests that the density of a taxon is limited by a habitat's productivity (Kaspari *et al.* 2000b). It is also used to describe variation in species richness over a large area (Rosenzweig 1995) and identify macroecological patterns. For different ant taxa habitats with a high net above-ground productivity (NAP), as in the tropics, have higher ant densities (colonies m^{-2}) and greater species richness than low NAP temperate habitats (Kaspari *et al.* 2000b) (Fig. 4.1).

These correlations are dependent on the scale of observation and generally decrease with increasing scale (plot to habitat). However, the mechanisms underlying these patterns are not clear. Other factors like habitat/landscape

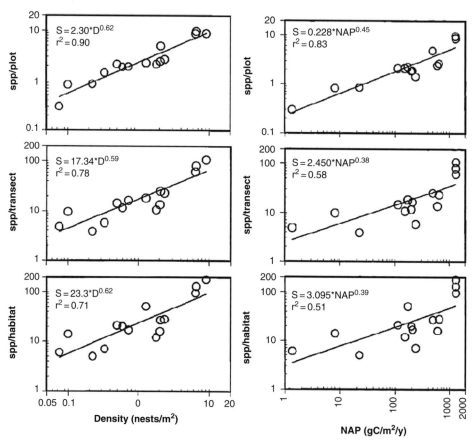

Fig. 4.1. Relationships between species richness of ground-nesting ants and nest density, and net above-ground productivity (NAP) at three spatial scales. Species richness is positively correlated with density and NAP, respectively, but the explanatory power of these predictive variables decreases with increasing scale from plot to habitat. (After Kaspari *et al.* 2000b.)

heterogeneity (Braschler *et al.* 2004), constraints on dispersal, competition, density dependence or dominance structure may frequently reduce diversity below the maximum set by NAP. In addition, worker size (Cushman *et al.* 1993) and colony size (Kaspari and Vargo 1995) both tend to decrease towards the tropics and large colony size commonly corresponds with small body size, and low per capita productivity (Karsai and Wenzel 1998). This implies that smaller ants with their high surface area to volume ratio may lose proportionally more energy as metabolic heat than larger individuals. Most likely this imposes severe constraints on small species, because they need to find the energy to maintain their high metabolism and foraging activities, but due to their smallness they might be raided by dominant ant species. Colony size of ants of same and different species may vary by several orders of magnitude and, as a consequence, energy requirements may vary accordingly. As indicated above, high densities of ants are unlikely to be sustainable if they are predominantly predators on other insects. Therefore, acquiring resources other than those from secondary consumers must be of primary importance for many species of ants and their insect partners. The organization of a colony might be monogynous or polygynous, possibly depending on habitat type (Seppa *et al.* 1995, Punttila 1996). Polygyny also affects the degree of relatedness of the individuals in a colony, which in turn could affect foraging behaviour and dispersal.

4.2.2 Aphids

Most aphids are usually short lived and, depending on species, produce 4–10 generations during a year. Each aphid can produce more than 30 offspring during its entire lifetime (iteroparous reproduction). There are basically two types of life cycle depending on whether an aphid spends all its life on a single host species (autoecious) or on different, often unrelated hosts (heteroecious), which requires that they produce winged morphs in spring that migrate to the secondary host and winged morphs in autumn that migrate back to the primary host (Fig. 4.2).

Briefly, and with a minimum of jargon, the life cycle starts in early spring with an overwintering egg giving rise to a fundatrix (stem mother). All subsequent generations are produced parthenogenetically; that is they are genetically identical in spite of the fact that they might be morphologically different, for example winged or unwinged. Offspring are produced viviparously. In autumn males are produced, which mate with females that lay overwintering eggs on branches or leaves close to the ground (Moran 1992, Dixon 1998). This cyclical parthenogenesis is a key feature of aphid reproductive biology, which

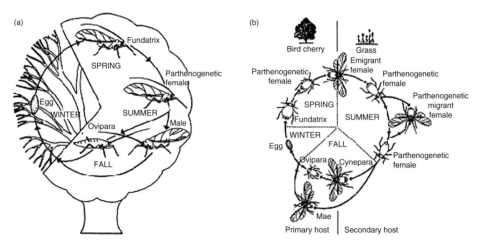

Fig. 4.2. Simplified life cycles of aphids. (a) Life cycle of the autoecious sycamore aphid and (b) that of the host-alternating (heteroecious) bird cherry-oat aphid. A wingless fundatrix gives birth to offspring that moult four times before reaching adulthood. Several generations might be produced during summer, which either possess wings (migrants) or lack wings. Two obligate migration periods, from a long-lived primary host to a short-lived secondary host, and back again, occur in heteroecious species. (After Dixon 1998. Copyright A. F. G. Dixon 1973, reproduced by permission of Edward Arnold (Publishers) Ltd.)

evolved early in the history of this taxon (Dixon 1998). Although the rates of increase of aphid populations are staggering, the life of an adult aphid is usually very short, often not exceeding a week or two. All developmental stages of myrmecophile species might be ant attended.

During larval growth resources are simultaneously directed to growth and reproduction, with daughters and grand-daughters developing in a mother. This 'telescoping' of generations (Dixon 1998) severely constrains migration to new host plants because the time they can spend without feeding is very short. Most aphids are highly host specific (even heteroecious species), which means that they can only live on a very narrow range of often closely related hosts. Some economically important species, however, infest a relatively wide range of host plants, but few aphid species are polyphagous (Kundu and Dixon 1995, Dixon 1998).

In the aphid–ant relationship, aphids, which have developed associations with ants, appear to have undergone little morphological change compared with other ant-attended species. Only cornicle length is reported to be reduced in ant-attended *Macrosiphini* (Mondor *et al.* 2002), but no special structures are developed. The suggestion that the rear of an aphid excreting honeydew (trophobiosis) resembles the front of an ant, and thus serves to appease ants (Kloft 1959), lacks empirical and theoretical support. While ant-attended lycaenid larvae have developed a harder cuticle, up to 20 times thicker than that of larvae not attended (Malicky 1970, Pierce *et al.* 2002), no such features are reported for ant-attended aphids. Some aphid species like *Monaphis antennata* on birch have particularly hard cuticles, which could be regarded as a defence against ant bites but they are not ant attended. It makes sense, however, because there are a number of species of aphids on birch that are closely attended by ants. In order not to fall victim to foraging ants and predators their dome-shaped body (Hopkins and Dixon 1997) and hard cuticle may make them more resistant to ant attacks.

4.2.3 Coccids

Scale insects are classified into three groups: (1) armoured scales (Diaspididae, about 2400 species), (2) soft scales (Coccidae, 1000 species), and (3) mealybugs (Pseudococcidae, 2000 species). Scale insects occur worldwide and like aphids are of economic importance because sap feeding may negatively affect plant vigour through the accumulation of sticky honeydew, which facilitates growth of sooty moulds, transmission of viruses or the stunting of new growth. The life cycle of scale insects starts with crawlers, which hatch from eggs laid under the body of adult females (Fig. 4.3). On hatching, they crawl until they find and attach themselves to a suitable place to feed, which is along leaf veins or new shoots. Crawlers can also disperse a long distance (up to several kilometres) either by wind or attached to birds or other insects (phoresy) (Greathead 1990). After completing three (females) or four (male) nymphal stages they become adults. On moulting to the adult stage, female scales remain attached for the rest of their life to a specific spot; males, however, pupate and eventually emerge as tiny fly-like, winged adults, which are rarely seen. An adult female scale can live for months to several years and produce several hundred eggs during her lifetime, which are laid either in a cavity beneath her body or in a waxy covering attached to her body. Several modifications are possible. Typically, most scale insects produce just one generation per year, especially in temperate regions. In contrast to some aphids and adelgids, host alternation is unknown in coccids.

Most scale species reproduce sexually but several different kinds of parthenogenesis are also known in scales (Miller and Kosztarab 1979), which suggests the multiple evolution of this kind of reproduction. The reason for this might be that the short-lived male adults are not always available and the flightless and sedentary females are restricted in their ability to move and seek sexual partners. This does not explain why males are short lived in the first place, but a proximate reason might be that investment into exclusively female

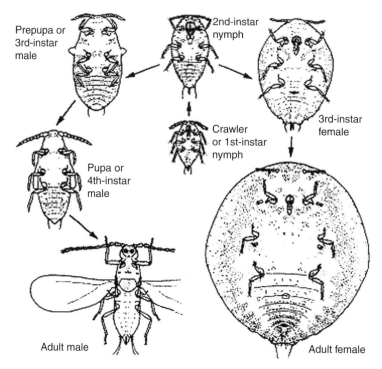

Fig. 4.3. Life cycle of scale insects. The crawler stage is the only mobile female stage and is able to disperse great distances. Because females have one nymphal stage less compared with males they are believed to be neotenous, while males have a holometabolic life cycle. (After Gullan and Martin 2003.)

offspring allows higher multiplication rates and quicker exploitation of ephemeral resources like the sap of a host plant. For example, Roff (1990) suggests that loss of flight is favoured in females because it permits greater allocation of resources to egg production, but that flight is retained in males because it increases the probability of finding a mate.

Neoteny shortens female development but results in the sexes maturing asynchronously. One way out of this dilemma is selection for smaller males, which do not feed and have shorter nymphal developmental times, which is the case for many male scales. In any case, a combination of features like sedentary adult females, low dispersal abilities of crawlers, parthenogenetic mode of reproduction and weakly flying adult males mating locally (Hanks and Denno 1993) may promote the formation of genetically distinct populations on patches of host plants. This might also affect the associations with ants and the frequency with which different levels of association develop. In this respect scale insects are closer to aphids than to membracids or lycaenids, in spite of the fact that the females are relatively long lived.

The following is restricted to a discussion of soft scale insects, which show the closest associations with ants (armoured scales are not ant attended). Soft scales, unlike armoured scales, do not secrete a protective scale-like covering. Soft scales typically secret a waxy, cottony or powdery cover. They are not flattened like the armoured scales but often hemispherical in shape, some almost globular at maturity. Both sexes are more mobile and move about for longer than armoured scales. During larval growth some may even move out onto the leaves as crawlers and return to the twigs in autumn before leaf fall. Eventually, they become stationary, feed, mate and lay eggs. Just like aphids, soft scales secrete copious amounts of sugary rich honeydew. Coccids even seem to have evolved a larger diversity of endosymbiont associations than aphids, which probably reflects their more diverse feeding strategies (Tremblay 1989). For example, parenchyma feeders might be less dependent on endosymbionts than xylem or phloem feeders because these plant storage cells contain more nutrients (Raven 1983). The change in feeding ecology during the evolution of coccids resulted in the complete absence or secondary loss of endosymbionts in several coccid groups (Tremblay 1989). Soft scales generally have one generation per year.

Just as in aphids the main benefit of ant attendance for coccids is protection against natural enemies and probably hygienic services (Bartlett 1961, Way 1963, Bach 1991), but some species do equally well without ants (Hill and Blackmore 1980). However, there is virtually no information on fitness benefits or costs to coccids of associations with ants and often it cannot be distinguished whether the benefit is an actual increase in the rate of reproduction, that is, a change in a life-history trait, or simply protection against natural enemies (decrease in mortality rates). It is important to know this because it would enable us to determine the relative benefits of associating with ants, and under what environmental conditions the options become advantageous. Of the three insect taxa included in our survey of the insect partners of ants, coccids are the least studied from a cost–benefit perspective and associated fitness consequences of ant attendance. Just for illustrative purposes: on doing a literature search for the key words 'coccid* and fitness' and 'aphid* and fitness' in the ISI Web of Science database there were 12 hits for scale insects but 227 for aphids. This is of course only a weak indicator of the quality of the research done on each group but it does suggest that there is a better understanding of the key life-history features of aphids than of coccids.

Clearly benefits vary with the identity of the ant species. For example, in a field experiment carried out in Papua New Guinea it was found that coccids attended by relatively inoffensive ants suffered higher incidences of parasitism than coccids attended by aggressive ants (Buckley and Gullan 1991) (Table 4.2).

Table 4.2. *Incidence of parasitism of coccids attended by ants showing different degrees of aggressiveness. The incidence of parasitism is significantly higher when the coccids are attended by inoffensive ant species (rank 1,2) compared with more aggressive ants (rank 3,4)*

Attendant ant		Number of tended coccid populations in which proportion of individuals parasitized is:		
Genus	Aggressiveness rank	0–10%	>15%	Total
Tapinoma	1		1	1
Iridomyrmex	2		2	2
Oecophylla	3	7		7
Solenopsis	4	1		1
Total		8	3	11

Source: After Buckley and Gullan (1991).

Ant exclusion experiments show similar results with those colonies from which ants were excluded, surviving for significantly shorter periods than attended colonies.

Although there is no general accepted measure of aggressiveness in ants a crude classification based on behaviour like biting, stinging, spraying of formic acid or mass recruitment is robust enough to provide some idea of the association between aggressiveness of attending ants and rates of parasitism. The percentage of coccids parasitized ranged from 0 to 95% in the different populations. Coccids attended by the relatively inoffensive ants *Tapinoma* spp. or *Iridomyrmex* spp. all show at least 15% parasitism, while for those attended by the more aggressive *Oecophylla* and *Solenopsis* species it never exceeded 10% (Buckley and Gullan 1991). So the important point is that being dependent on the services of ants might have a very different outcome depending on the dominance status and aggressiveness of the ants. Being attended by subdominant ants could impose severe costs, as the quality of the services received by the partners of these ants is highly uncertain. As a consequence, the distribution of dominant, codominant or subdominant ant species may affect the size and composition of a coccid community.

There is relatively little information on how the quality of the honeydew excreted by coccids affects the degree of attendance or the association with particular ant species. There is some indication that workers of *Linepithema humile* prefer the honeydew of *Coccus hesperidum* over that of other coccids like *C. pseudomagnoliarum* (Ewart and Metcalf 1956) suggesting some kind of preference hierarchy, as reported for aphids. It is possible that secondary compounds, which are also present in coccid honeydew, might affect these ant–homopteran associations (Molyneux *et al.* 1990) but specific information on coccids is not available. In particular, in the literature on coccid–ant interactions the idea that plant quality determines honeydew quality of the coccids, and cascades through ant–coccid mutualisms, still prevails. Information on the active modification of the quantity of honeydew offered to ants, composition and varying modes of excretion, temporal variability in honeydew composition, or effects of alternative sugar resources on coccid population dynamics and community structure is still missing, but essential for understanding the selection pressures operating in mutually exploitative partnerships.

4.2.4 *Membracids*

Females of many subsocial treehoppers are semelparous; that is, they produce a clutch of eggs during a short reproductive period of not longer than 2–3 days. They deposit a cluster of eggs on a shoot of a plant (Fig. 4.4) and females often stand on top of their eggs and guard them and their emerging offspring (Wood 1993). Over the next couple of days to months females guard their eggs and nymphs (Fig. 4.5) almost throughout their development. The adult females display defensive behaviour when approached by a potential enemy. The behavioural repertoire includes kicking, wing fanning and even communication with their offspring (Cocroft 1996, 1999). However, it is not clear whether the co-ordinated production of signals by the nymphs is co-operative or whether they compete for the mother's proximity, because parental anti-predator behaviour is clearly a limited resource for the offspring. Cocroft (2002) showed that in the thornbug treehopper, *Umbonia crassicornis*, the presence of the mother, the relative position of the nymphs and distance from the female clearly determined the mortality of the offspring. The individuals feeding on the edges of colonies and furthest away from the mother suffered a higher mortality from attack by predatory wasps.

Like many other homopterans, treehoppers engage in mutualisms with ants, provide honeydew for the ants and receive some form of protection from natural enemies. In combination with maternal care this could increase the fitness of offspring provided the ants can be effectively manipulated to benefit the treehoppers. There is substantial evidence that ant attendance considerably increases survival of nymphs but the picture might be more complex than thought. For example, in the treehopper *Publilia modesta*, ant tending increases nymphal survival, while maternal care in the absence of ants is relatively

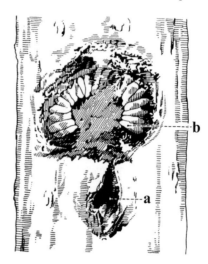

Fig. 4.4. Females of *Ceresa bubalus* deposit their eggs in a chamber in the bark of the host plant, which they excavate with their ovipositor. a marks entrance to the chamber; b shows bark removed to expose the eggs. (After Grassé 1951.)

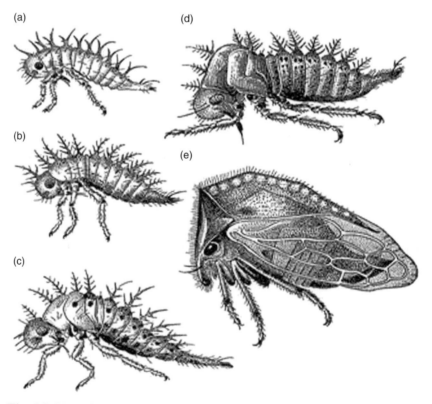

Fig. 4.5. Nymphs (a–d) and adult (e) of *Ceresa bubalus*. (After Grassé 1951.)

ineffective or even negative. A significant ant × maternal care interaction indicates that maternal care increases the number of nymphs surviving to adulthood, but only in the presence of ants (Billick *et al.* 2001). The main effect of the mother appeared to be to attract ants and increase the per capita tending rates of the ants, especially during the first weeks of nymphal development. Mothers achieve this by providing an abundance of honeydew for the ants. However, it is unknown whether the increase in honeydew excretion is accompanied by a change in honeydew composition. Thus, maternal care is an elaborate strategy for attracting ants early in nymphal development and until the nymphs are large and produce sufficient honeydew to be attractive to ants, when the females leave their offspring and produce another clutch on another shoot. If honeydew production also incurs costs in membracids, as in aphids (Yao *et al.* 2000) and lycaenids (Pierce *et al.* 1987, Robbins 1991), then it is not the offspring that incur the costs. However, there might be future costs for females.

Cost–benefit trade-offs in female brood care have been demonstrated in a closely related species of membracid, *Publilia concava*. In this species females also commonly guard their eggs. Given that brood care can be turned over to attending ants there is a trade-off between the period of time for which the eggs are protected and initiation of a second clutch. Long periods spent guarding could negatively affect the number of clutches, as was shown by Del-Claro and Oliveira (2000). These fecundity costs are likely to be of more importance in species that lay several clutches. In *P. concava*, the period of time a female stays with her brood clearly affects hatching success (Zink 2003) (Fig. 4.6). Although the variation in hatching success varies considerably there nevertheless is a significant linear relationship with female guarding time. For example, a female that stays for 30 days doubles its hatching success relative to one that immediately abandons its brood (0.3 nymphs/egg versus 0.15 nymphs/egg).

Early abandonment of eggs allows females to lay more clutches (larger number of future clutches; Fig. 4.7a). In particular, females that abandon broods within 0–4 days initiated a significantly larger number of secondary broods compared with those that remained. Abandoning the initial brood later had no effect on number of future broods. Females that abandoned their broods early produced a second full-sized clutch of eggs (on average 6–7 mm), which is significantly larger than that produced by later abandoning females (Fig. 4.7b). The difference between remaining and abandoning females was apparent for females that abandoned their brood within 10 days (Fig. 4.7b). In addition, females that abandoned their brood guarded future broods for a longer period than those that guarded their first clutch for 11 days or longer (Fig. 4.7c).

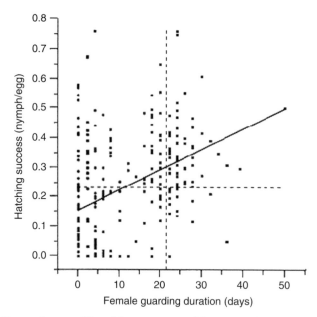

Fig. 4.6. Dependence of hatching success of broods of *Publilia concava* on female guarding time. Dashed lines give the averages for both variables. $r = 0.42$, $n = 280$, $P < 0.0001$. (After Zink 2003.)

Adding up the total sizes of all the egg masses produced over the lifetime of a female resulted in early abandoning females having higher lifetime fecundities. In order to compare the two strategies and correct for variable hatching dates Zink (2003) subtracted the departure date from the hatching date for each female and plotted the frequency distribution of these time differences for each female (Fig. 4.7d).

The frequency plot of these values showed a strong bimodal distribution, which suggests the existence of two strategies. Early departers quickly abandon their brood to initiate another one and late departers place less value on future broods but take care of the first brood (Fig. 4.7d). Interestingly, females that fall into these two groups have similar lifetime fitness. This suggests that the alternative tactics of abandoning and guarding might be maintained in populations or might change depending on environmental conditions. Intermediate strategies apparently are less frequent and possibly selected against because of a mismatch in life-history characters or external factors, which might be correlated with plant quality or seasonal variation in plant quality. For example, it is suggested that female care might be more appropriate when oviposition sites are of high quality, when predator pressure is high, or when there is a high incidence of ant attendance (Zink 2003). Under

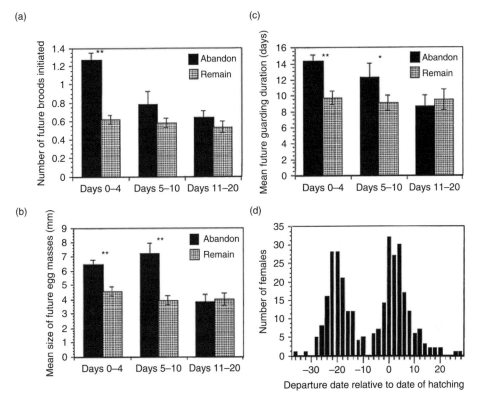

Fig. 4.7. Means (± 1 SE) of the future number of broods initiated (a), size of future broods (b), and future guarding duration (c) for females abandoning within 0–5 days or remaining for 6–10 or 11–20 days guarding their eggs. **$P < 0.0001$, *$P < 0.01$. (d) Distribution of departure dates of females from their first brood relative to the day of hatching. (Modified after Zink 2003.)

these conditions the benefits of remaining might be large and females can be expected to adopt a strategy of guarding until hatching. Conversely, under conditions in which the costs of remaining are large, females might be expected to adopt a strategy of brood abandonment. This might occur, if females have many more eggs to lay, if predator pressure is low, or if females have a limited amount of time to initiate a second clutch. Similar to *P. modesta*, females of *P. concava* have a higher nymphal survival when attended by ants. This again suggests that one (if not *the*) advantage of female guarding is that it attracts ants. Females of *P. concava* are also more reluctant to oviposit on plants without ants (Morales 2002), which makes sense if by turning over parental care to ants clutches can be abandoned earlier and another brood initiated. These are good examples of the trade-offs, costs and benefits associated with parental care, ant attendance and associated services in treehoppers. These associations are conditional and only exploited if they confer fitness benefits on the recipient.

Another aspect of ant–membracid mutualisms, which runs through this book, is density dependence, and potential costs and benefits when associated with ants. As in aphids, small colonies of membracids might not produce sufficient honeydew to be attractive for ants, while in large aggregations the ratio of the number of ant partners/ant might become too large for the ants to provide efficient protection services. In a field survey in southern Connecticut involving the ant *Formica obscuriventris* and the treehopper *Publilia concava* feeding on *Solidago altissima* it was shown that the difference in the number of nymphs surviving between tended and untended treehoppers was highest for small aggregations and decreased significantly as colony size increased (Morales 2000a). That is, small colonies benefited most. This could be a result of the recruitment behaviour of ants. While the total number of ants tending treehopper aggregations increased with *Publilia* colony size the per capita tending rates were higher in small colonies and decreased with growing aggregation size. Other studies, however, have found that intermediate and large aggregations of treehoppers benefit most (McEvoy 1979, Cushman and Whitham 1989). These different findings might be due to several factors. For example, ant recruitment patterns significantly change in response to the spatial configuration of ant nests and honeydew resources. Resources far away from ant colonies are likely to be deserted earlier than those close to (Taylor 1977, Sudd 1983). Alternative sugar resources might also lead to competition for mutualists. For example, Bristow (1984) showed that the benefits honeydew producers derive from attending ants are unequal and asymmetric, with aphids (*Aphis vernoninae*) benefiting more from associations with the ant *Tapinoma sessile* and the membracid *Publilia reticulata* than from associations with *Myrmica* species. Similarly, intraspecific competition for mutualism appears to affect the relative benefits derived from ant attendance (Cushman and Whitham 1991). However, evidence of competitive interactions for the services of ants is equivocal. Del-Claro and Oliveira (1993), for example, found no effect of alternative sugar resources, such as extrafloral nectaries, on the tendency of *Camponotus* ants to attend membracids. A possible explanation is that honeydew excretion was greater in the presence of alternative sugar resources, which also suggests an adaptive density dependent response of membracids aimed at securing the services of ants. Whatever the underlying mechanism, context dependent positive or negative density dependent benefits make associations between ants and their homopteran partners highly dynamic and unpredictable, and the partner needs a good repertoire of behavioural responses in order to cope with this problem.

4.2.5 *Lycaenids*

In lycaenids all the life stages might interact with ants (Fig. 4.8).

Although most species feed on living plant tissue, some are also entomophagous during the immature stages (Pierce *et al.* 2002). There are an estimated 6000 species of Lycaenidae, which accounts for about one third of all Papilionidae (Shields 1989). In contrast to other butterfly lineages, only in the two monophyletic sister taxa, the Lycaenidae and the Riodinidae, did an ability to form associations with ants evolve. Here only the lycaenids are considered because most of their associations with ants are well documented (DeVries 1991a, Fiedler 1991). For those species whose life history is fully known, about one third are closely associated with ants (obligate myrmecophiles), 45% show a more casual association (facultative myrmecophiles) and about a quarter have no association with ants (Table 4.3). Note, however, that obligate and facultative interactions include both mutualistic and parasitic species.

If only phytophagous larvae of the subfamily Lycaeninae are considered then about 27% show no association with ants, 60% are facultative myrmecophiles and 13% obligate myrmecophiles (Osborn and Jaffe 1997). That is,

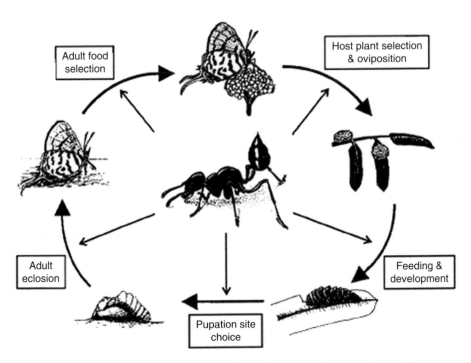

Fig. 4.8. Life cycle of lycaenid butterflies and potential interactions with ants. All life stages can interact with ants but most importantly the caterpillars. (After Fiedler 1997b.)

Table 4.3. *The number and relative frequency (in parentheses) of taxonomic groups (subfamilies and tribes) of lycaenids showing different degrees of associations with ants*

		Ant-association (%)		
Taxonomic group	N	Obligate	Facultative	None
Poritiinae	44	0	0	44(100)
Poritiini	1	0	0	1(100)
Pentilini	11	0	0	11(100)
Liptenini	32	0	0	32(100)
Miletinae	27	14(52)	0	13(48)
Liphyrini	5	4(80)	0	1(20)
Miletini	9	4(44)	0	5(56)
Spalgini	4	0	0	4(100)
Lachnocnemini	9	6(67)	0	3(33)
Curetinae	6	0	1(17)	5(83)
Lycaeninae	588	183(31)	299(51)	106(18)
Theclini	226	56(25)	117(52)	53(23)
Aphnaeini	95	92(97)	3(3)	0
Lycaenini	36	0	5(14)	31(86)
Polyommatini	231	35(15)	174(75)	22(10)
Total	665	197(30)	300(45)	168(25)

Source: After Pierce *et al.* (2002). Records compiled by Pierce *et al.* (2002) from numerous sources.

only a minority of species are closely associated with and their survival completely dependent on ants.

   Like all other partners of ants lycaenids manipulate the behaviour of ants by suppressing their aggressive behaviour towards them but exploiting it as a means of defence against natural enemies. This is achieved by regulating the quantities and quality of the nectar provided for ants (Agrawal and Fordyce 2000). For example, larvae of the obligate myrmecophile *Jalmenus evagoras* secrete more nectar when feeding alone than when feeding in small groups. In addition, larvae secrete less nectar when with a bigger more attractive larva than with a small larva (Axen and Pierce 1998) and different ant species trigger different secretion behaviour (Axen 2000). The ability to adapt secretion rates to the social environment and perceived predation risks indicates that gaining protection at a reduced cost is an important factor promoting aggregation, probably not only in myrmecophile lycaenids.

   Attending ants can significantly affect the fitness of lycaenid larvae. Fitness is often measured indirectly via larval or pupal weight, adult size or number of eggs laid. For example, laboratory experiments show that *J. evagoras* larvae attended by *Iridomyrmex* ants take almost five days less to develop compared with unattended individuals, when the total development takes about four weeks (Pierce *et al.* 1987). However, the pupae of tended females are 25% lighter than those of unattended females and eclosing adults are also smaller. Similarly, larvae of the neotropical *Arawacus lincoides* showed reduced growth when attended by *Ectatomma* ants (Robbins 1991). Because female size is often correlated with female fecundity this is likely to result in a significant fitness loss. Of course this fitness loss is compensated for if attended larvae suffer a lower mortality due to predation, as is frequently the case in lycaenids (Pierce and Easteal 1986, Pierce *et al.* 1987, Wagner 1993, Cushman *et al.* 1994, Fiedler *et al.* 1996, Seufert and Fiedler 1996a). Not all lycaenids respond to ant attendance in a similar, negative way. For the facultative myrmecophile lycaenid *Hemiargus isola*, attendance by three different ant species did not result in reduced larval size or female fecundity. In particular, tending by *Formica perpilosa* significantly enhanced larval growth resulting in a 20% increase in body mass compared with unattended conspecifics. In contrast, tending by *Dorymyrmex* spp. and *Forelius foetida* had no significant effect on larval mass. Clearly, tending has physiological costs as non-feeding tended larvae lost 69% more weight compared with non-feeding unattended larvae (Wagner 1993). Compared with the results reported for obligate myrmecophiles this is a somewhat surprising outcome because facultative myrmecophiles are not dependent on ants and it is difficult to understand why these species should not always feed at their maximum rates. A possible solution is that ant attendance allows them to feed without being interrupted for longer, which would suggest that they are more dependent on ants than classifying them as facultative myrmecophiles implies. A recent review, including data on more than 150 parasitic and phytophagous species of lycaenid, suggests that obligate myrmecophily often constrains growth compared with that achieved by facultatively associated or unassociated species (Elmes *et al.* 2001).

Food selection can be influenced by myrmecophily as adult females oviposit on plants where ants are present. This ant-dependent oviposition is reported for many species and seems to be most prevalent in certain taxa like the Aphnaenini (Pierce and Elgar 1985, Fiedler 1991). Often clutches of eggs laid by obligate myrmecophiles are larger than those of related species that are less closely associated with ants (Kitching 1981), which may be advantageous when attracting ants. Using ants as a cue for oviposition has an important consequence. If the presence of ants enhances the fitness of their offspring females might have to decide whether to lay eggs on a plant that is of superior quality or on any plant as long as ants are present. This might favour the development of a more polyphagous lifestyle than that of unattended or facultative myrmecophilous species. There is no good test of this idea but circumstantial evidence suggests that obligate myrmecophiles tend to use a broader spectrum of host plants than unattended and facultative myrmeco-philes do (Pierce and Elgar 1985, Fiedler 1994).

Because ant-associated lycaenids produce sugary and protein-rich secretions as a reward for ants it is also suggested that they are more likely to feed on legumes and other nitrogen-rich host plant species. This should result in the narrowing of the food plant spectrum. Again, there is some support for this as ant-attended lycaenids are significantly more likely to feed on Fabaceae and nutrient-rich inflorescences compared with unattended species (Fiedler 1995, Stadler *et al.* 2003). Fertilization experiments also indicate that larvae of *J. evagoras* feeding on nitrogen-rich fertilized plants attract significantly more ants than those feeding on unfertilized plants and adult females even seem to prefer to lay eggs on these high-quality plants (Baylis and Pierce 1991). However, whether this occurs generally is debated (Fiedler 1995) not least because of the limited number of experiments and poor knowledge of the phylogenetic relationships of many species and their relationship to their host plants. Other bottom-up effects, like secondary chemicals, might play an equally important role in influencing ant–lycaenid associations (Fiedler 1996).

All mutualisms between ants and lycaenids are subject to the influence of alternative sugar resources, which can be other homopterans or extrafloral nectaries (EFNs). For example, EFNs are more abundant in tropical than in temperate plant communities (for reviews see Pemberton 1998, Oliveira and Freitas 2004), which might suggest that in association with the higher ant diversity in the tropics ant–EFN relationships are more likely to affect ant–myrmecophile interactions there and specific adaptations are necessary for survival in an ant-dominated environment. In a survey of woody plant species in the Brazilian Cerrado 15–26% of all species bear EFNs. Nevertheless, a number of species of Lepidoptera co-occur on ant-visited plants; some are ant attended, others are not. In order to circumvent attack by ants these species have evolved a rich set of behavioural traits. For example, larvae of the non-myrmecophile nymphalid *Eunica bechina* eat the young leaves of the ant-visited EFN shrub *Caryocar brasiliense*. The caterpillars construct needle-like 'frass chains' at the tips of leaves, where they retreat to escape patrolling ants (Oliveira and Freitas 2004). (The use of the term 'frass' in English is based on a mistranslation of the German word *Fraß*, which correctly translated is the 'amount eaten'.) The pile of frass produced by a caterpillar is climbed on to escape patrolling ants. Older larvae can also defend themselves by regurgitating or bleeding upon seizure by worker ants, which inhibits further attacks. Still another option available to larvae is dropping off a leaf and hanging by a silk thread. The success of different ant species in capturing *Eunica* larvae clearly varies (Freitas and Oliveira 1992), which again indicates the conditional dependence of interactions. Taken together, this arsenal of defence enables larvae of this unattended species to survive even on highly ant visited EFN-bearing host plants. It also shows that mutualistic associations with ants are just one way of exploiting host plants that offer rewards to ants.

Generally, there is a broad overlap between ant communities collecting nectar at EFNs and those visiting lycaenid larvae (DeVries 1991b). This underscores the fact that many facultative lycaenid–ant interactions are characterized by a high degree of opportunism and low degree of specificity. The services of ants are just another resource that may be exploited if it increases fitness. But dependence on the services of ants introduces constraints, uncertainties and exploitation costs, especially if cheaters or alternative resources are attractive to ants and must be competed with, if different ant species provide different degrees of benefit (Jordano and Thomas 1992), or if the spatial context of such interactions is important (Jordano *et al.* 1992).

4.3 Emerging patterns in the distribution of outcomes

Most partners of ants belong to two insect orders, Homoptera and Lepidoptera, which are primarily phytophagous. The basis of these associations is that partners of ants produce some sort of food, while ants provide protection or hygienic services. Food for ants can be excreta, which originally required little modification as ants might have collected it from the surfaces of leaves. Myrmecophilous lycaenid larvae, however, produce nectar from specialized glands, which probably represent a more costly investment than that involved in modifying the quality or quantity of honeydew. It is tempting to look for broad patterns in terms of what phylogenetic, genetic or ecological features facilitate ant attendance in these groups and each of these factors will be addressed in turn. However, there is also the risk that broad generalizations across different taxa, such as aphids, coccids, membracids and butterflies, may be grossly misleading. We believe it is equally instructive to look for differences within taxonomic lineages and ask why certain species evolved associations with ants while closely related species did not. This raises an interesting question: How important are ants for the evolution and diversification of partners of ants? Do ants serve as templates, for example for butterfly or aphid diversification (Pierce *et al.* 2002), or has the role of myrmecophily in the evolution of species diversity of lycaenid butterflies (and potentially other insect partners of ants) been overestimated in the past

(Fiedler 1997b)? If there is an answer to these questions for lycaenids, which make the highest investment in ant attendance, then such an answer might similarly apply to homopterans, whose investment in ant attendance is smaller. The answer, however, should not be based only on phylogenetic criteria but must recognize the ecological context as shown below.

First we review the phylogenetic information on ant attendance.

4.3.1 *Phylogeny*

Although there are some geographic and taxonomic clusters where the incidence of ant attendance is higher in lycaenids and other partners of ants, and in spite of the fact that a detailed phylogenetic analysis of the frequencies of ant attendance is still missing, preliminary analyses indicate that in most taxa of lycaenids, coccids, membracids and aphids ant attendance either never developed, developed, or developed and was then lost (Pierce *et al.* 2002, Shingleton and Stern 2003, Stadler *et al.* 2003). This suggests that mutualistic relationships with ants are rather labile and can be given up if the associated fitness costs become larger than the benefits. Surprisingly so, this is even the case for lycaenids, where associations with ants could not evolve as a by-product, as in sap-feeding insects. Clearly, there is some phylogenetic conservatism involved with respect to ant attendance. For example, within certain lycaenid genera, such as *Ogyris* or *Jalmenus*, all species are associated with ants. In the small group of Australian lycaenids of the genus *Jalmenus* all species primarily interact with the ant *Iridomyrmex purpureus* (Eastwood and Fraser 1999). This mutualism with ants could have a strong effect on population structure. Being dependent on ants and on specific host plants can lead to extreme habitat fragmentation and restriction to patches where conditions are good. As a consequence, it is argued that site fidelity, assortative mating, smaller population sizes and genetic isolation of populations might facilitate diversification (Atsatt 1981, Smiley *et al.* 1988, Jordano and Thomas 1992, Costa *et al.* 1996). However, whether this is generally true for phytophagous, free-living lycaenid larvae is less certain (Nice *et al.* 2002) and current evidence even for obligate myrmecophiles like *J. evagoras* does not support the idea that diversity is driven by their associations with ants (Costa *et al.* 1996).

Given that the total number of obligate myrmecophiles in the different taxa described above is somewhere between 10% and 20% it is unlikely that ant attendance is a radiation platform for homopteran and lycaenid evolution. Most associations are facultative, indicating some kind of tolerance, appeasement strategy or unspecific mutual exploitation. Even among the obligate myrmecophiles and well-studied species very few cases (if any) have convincingly shown that myrmecophily is the main cause of speciation events. Myrmecophily is present in all taxa, with lycaenids showing more conservatism at the species level than homopterans, where species within the same genus have developed quite different levels of associations with ants (Stadler *et al.* 2003).

4.3.2 Genetics

There is currently no evidence that ants restrict gene flow more so than ecological factors, such as host plant distribution, patchiness, or secondary plant compounds. The degree of relatedness between individuals in a colony attended by ants might, however, influence the relative costs and benefits of myrmecophily for a genetical individual (the genet). For a fast reproducing clone, consisting of identical individuals, losing a few individuals to predatory ants might not affect the fitness of the genetical or evolutionary individual (Janzen 1977) in a significant way, whereas, for example, the offspring of a lycaenid female are all unique genetically and losing an individual to ants means the loss of a unique gene combination.

4.3.3 Ecology

Given that there is no detailed phylogeny of lycaenids and homopterans across different taxonomic levels and preliminary evidence suggests that myrmecophily is a labile trait rather than a phylogenetically constrained feature, it is probably more interesting to ask what ecological features promote antagonism or mutualism in different groups of ant partners. Phylogenetic information is of limited use when attempting to account for the existence of different levels of myrmecophily in closely related species, which have the same ecological requirements. In all groups of insect partners there are cases where closely related species co-occur on the same host plant but maintain different levels of associations with ants. For example, the Malaysian legume tree, *Saraca thaipingenses*, is used by 11 species of lycaenids that feed on the young foliage and inflorescences (Seufert and Fiedler 1996b). Three of them have been studied in detail (*Drupadia theda*, *D. ravindra*, *Cheritra freja*). While they have similar ecological requirements with regard to larval host plant their levels of association with ants differ markedly. *Drupadia theda* caterpillars are obligatorily associated with two *Crematogaster* species, those of *D. ravindra* facultatively with several ant species and those of *C. freja* are not attended. Many similar examples can also be cited for aphids and membracids. For example, on tansy (*Tanacetum vulgare*) a dozen aphid species co-occur but only one is an obligate myrmecophile, while the others are either facultative myrmecophiles or not associated with ants. The reasons why these different strategies coexist are manifold and it is highly likely that each association is advantageous at a particular time or place (Stadler 2004). The relative benefits of these strategies vary and it is proposed that differences exist in the relative abilities of the species to colonize new host plants or to compete with each other in a mosaic of different environments (colonization–competition trade-offs) at local and regional scales. More detailed examples of associations with Homoptera will be provided in the next chapter.

Further similarities, which affect all partners of ants, are that associations with ants are conditional and unpredictable. Several studies are highlighted that report a significant spatial and temporal variation in the interactions between ants and their insect partners and that the species of ant affects the relative benefit for the myrmecophile (Buckley and Gullan 1991, Peterson 1995, Fiedler 1997b, Fraser *et al.* 2001, Braschler and Baur 2003, Braschler and Baur 2005). All closely associated partners of ants are likely to be exposed to the same selection pressures. For example, they might face a trade-off between polyphagy and myrmecophily. Ants, or suitable species of ants, which provide the best services, might not be available in each habitat. As a consequence, a female might have to decide whether to oviposit on plants that are less suitable for her offspring's growth and development, but where ants are available, or on high-quality plants without ants. Hierarchical and temporal preferences for different species of ants (Fraser *et al.* 2002) might be a direct consequence of these trade-offs. The fact that many phytophagous insects are specialists rather then generalists probably indicates the difficulty of exploiting the services of ants.

Irrespective of the species all mutualisms are subject to exploitation. Many cases are known of specialist predators and parasitoids that specifically exploit the associations between ants and their myrmecophiles (Völkl 1992, Seufert and Fiedler 1999, Kaneko 2002). This might be an especially rewarding strategy because partners of ants are usually clumped, relatively immobile, provide ample resources for the exploiter and the ant might also provide enemy free space for the exploiter. From the point of view of ant partners an increasing predator or parasitoid pressure from such specialists might facilitate the secondary loss of myrmecophily. Currently, however, there is no way of quantifying how often these losses have occurred because of the paucity of phylogenetic data at a variety of taxonomic levels.

All these mutualisms are subject to bottom-up and top-down effects at local and regional scales (Cushman 1991, Denno *et al.* 2002, Billick and Tonkel 2003). This includes habitat and plant quality, habitat fragmentation, which affects dispersal success, the species of ants present in particular habitats and the varying degrees of services provided, competition between partners of ants and extrafloral nectaries for the services of ants, which might lead to temporal preference hierarchies for each mutualist. This includes a clear cost–benefit perspective, which has been most successfully adopted for aphids, membracids and lycaenids but less so for coccids. For example, attracting and successfully competing for ants requires the production of more or better nectar/honeydew than the competitor or the EFNs, which is an investment that will appear on the fitness balance sheet together with the potential pay-offs. Generally, for facultative myrmecophiles it is easier to observe physiological costs than for obligate myrmecophiles. This is because they should always feed at their maximum rate irrespective of whether attended or not. The provision of ants with food should thus deprive them of some food, which would otherwise have been used for reproduction. The fact that all partners of ants are able to adapt their secretion rates to tending levels strongly argues for physiological costs of attendance.

Still another similarity of many of the associations described above is the density dependent nature of costs and benefits (Addicott 1979, Morales 2000a). Costs of ant attendance might decrease with the size of the attended colony because each individual has to pay less but simultaneously benefits less with increasing population size of the myrmecophiles because the per capita protection service declines. Allee-like functions are equally conceivable. That is, small and large colonies might benefit less than intermediate sized colonies because small colonies are unable to produce enough sugary solution, while large colonies suffer the negative effects of deteriorating plant quality and increase in competition. However, the period for which the optimum colony size lasts varies for the different partners of ants. For example, aphids are parthenogenetic and often show an exponential increase in numbers, which might quickly exceed the tending capacities of subdominant species of ants. In membracids and lycaenids, where the females produce a clutch of eggs, honeydew/nectar production increases only as a consequence of the offspring growing and therefore honeydew production is unlikely to exceed the handling capacities of the ants and the optimal group size either lasts longer than it does in aphids, or is never achieved.

IN SUMMARY, although the protection hypothesis is well established for all groups of myrmecophiles its very absoluteness masks the context-dependent nature of relationships with ants, which range from antagonism to mutualism. The relative predation risk and the genetic structure of the populations determine the impact of bottom-up and top-down forces and ants are just one component of the forces that affect fitness. This has been shown for membracids (Denno *et al.* 2003), aphids (Stadler 2004) and lycaenids (Fiedler 1995, Pierce *et al.* 2002). Although morphological and behavioural adaptations of the partners of ants are frequently reported (Malicky 1969, Gullan and

Kosztarab 1997, Mondor *et al.* 2002, Pierce *et al.* 2002) ant attendance is unpredictable as it is not phylogenetically determined and can only be understood by considering the combined effects of life history, populations and communities in a special setting (see Chapter 6). Before proceeding with the analysis of multilevel interactions a closer examination of the associations between aphids and ants is provided, because they provide a wealth of information on all these aspects. The view adopted will be that the effects of ants on aphids and of aphids on ants are of equal importance in shaping the antagonistic–mutualistic continuum.

5

A special case: aphids and ants

5.1 Features associated with ant attendance

Phloem sap is a poor diet because of its low nitrogen concentration, unbalanced composition, and temporal variability in quality (e.g. nitrogen or secondary metabolite content). Nitrogen is mostly present in low concentrations usually ranging between 50 and 300 mM (0.8–4.5% w/v) (Mittler 1958, Ashford *et al*. 2000, Sandström and Moran 2001). Intracellular aphid symbionts (*Buchneria*) provide their host with amino acids, which are otherwise in short supply in the phloem sap, and the symbiosis between aphids and bacteria is considered as essential for utilizing phloem sap (Douglas 1998). Adaptations to utilize this resource might offer opportunities and cause problems with respect to mutualistic interactions with ants. For example, different ant species and different groups within an ant colony have different nutritional needs at different times of the year. Workers rely on carbohydrates for their energy needs during foraging, whereas larvae require mainly nitrogenous food for growth. Like phloem sap, honeydew is a nitrogen-poor diet and, as a consequence, ants must be able to adapt to this resource. For example, some ant species harbour micro-organisms in their digestive tract, probably to supplement the liquid diet with essential amino acids and other nutrients (Roche and Wheeler 1997).

A large number of adaptations have been shown (some of which were described in previous chapters) to influence to some extent the strength of the interactions between ants and aphids, which range from positive to negative. An overview of the physiological, ecological and evolutionary traits that might favour either mutualistic or antagonistic interactions between aphids and ants is given in Fig. 5.1. These features should be viewed from a population biology perspective and include competition–colonization trade-offs, density dependence, relative frequencies of high and low quality patches and top-down/bottom-up

Features associated with:

	Mutualism	Antagonism
Physiological	(a) • High growth rates – at least temporarily • Ability to make the sugar composition of honeydew more suitable for ants • Honeydew is a waste product needing little further investment • Energy source for high activity 'tempo' foragers	(d) • Cost of producing high quality/large amounts of honeydew • Changing nutritional requirements of ants during their life cycle e.g. need for less honeydew
Ecological	(b) • Hygienic services • Protection by ants • Habitat fragmentation • Abundance of aphids • Distribution of aphids • High quality of host • Gregariousness	(e) • Predation by ants • Competition for mutualists • Fitness costs • Low predictability of C-resource • Alternative sugar sources (e.g. EFNs) • Chemical defence of plants affects honeydew and ants
Evolutionary	(c) • Proventriculus for storing honeydew • Extensible gaster further facilitates the storage of honeydew • Overcome initial defence or aggressiveness	(f) • Low predator/parasitoid pressure (association costs) • Exploiters of mutualism (e.g. other aphids) • Exploiters of mutualism (e.g. specialized parasitoids, predators)

Fig. 5.1. Factors that influence the strength and direction of the associations between aphids and ants. Temporal or spatial aspects of these factors are rarely documented. (Modified after Stadler and Dixon 2005.)

effects for both aphids and ants. For example, the benefits of hygienic services often associated with ant attendance only become relevant if the population size of the honeydew producers is large and when aphids feed in aggregations. Feeding in dense clusters, however, is likely to affect plant quality and eventually the need to disperse to new hosts. Similarly, changing frequencies of certain patch types over the growing season might influence the relative benefits of mutualism between aphids and ants (Section 5.2). A mutualistic relationship between aphids and ants can only be expected if both partners manage to exploit each other in a way that enables each partner to achieve higher population growth rates and larger colony sizes (Fig. 5.1a). For example, by modifying honeydew sugar composition and concentration it may temporarily become more attractive to ants, which in turn must be able to modify their foraging behaviour so that they can effectively harvest this energy-rich resource.

Other ecological effects (or ecological by-products of the relationship) are well documented. For example, hygienic services, like the collection of sugary excreta and protection from fungal infections (Fig. 5.1b), are also thought to favour mutualism. However, these are likely to be secondary because unattended aphids effectively dispose of their honeydew and are apparently not so prone to fungal infections associated with honeydew. The fragmentation of habitats, their complexity and plant-related factors, such as abundance, wide distribution and quality, are thought to facilitate mutualism because these attributes result in an increase in aphid abundance, which increases the probability of their being encountered by ants. Put another way, obligate mutualisms are unlikely to evolve between rare species. Lastly, morphological and behavioural adaptations that enable ants to collect large amounts of liquid food and feed it to their non-foraging nest mates and larvae are likely to facilitate the successful establishment of mutualistic relationships (Fig. 5.1c). Selection for an antagonistic relationship is likely if the costs associated with ant attendance – such as the production of high-quality honeydew, the absence of ant partners in many habitats or inaccessibility of suitable hosts – are high (Fig. 5.1d). If plants compete with aphids for the services of mutualists (for example via extrafloral nectaries, EFNs), or if the chemical protection of the host plant affects the ants (e.g. honeydew containing secondary plant metabolites), an intimate relationship is less likely to develop (Fig. 5.1e). Similarly, if mortality due to natural enemies is low or if specialized predators or parasitoids exploit ant-attended aphids (Völkl 1992, Kaneko 2002) (Fig. 5.1f), costs might outweigh benefits. Essentially, these features operate at different levels of organization, temporal and spatial scales and will be discussed in more detail in Chapter 6.

5.2 Cost–benefit perspective

All interactions between ants and aphids, when framed within the competition-colonization trade-off, entail costs and benefits for both partners. The terms costs and benefits are rather vague and do not portray the specific fitness gains or losses for clonal or highly related organisms, but give some indication of the constraints in these associations. Benefits for one partner may entail costs for the other. Therefore, the tension within mutualistic systems and the relative magnitude of the resulting conflict/shared interests determines whether a relationship is positive or negative. At the risk of oversimplification, ants are very active and many forage over great distances, so energy for foraging is likely to be an important limiting factor. Aphids have to process large quantities of phloem sap to sustain their high growth rates, so honeydew is often likely to be abundant and available for fuelling ant foraging. However, because phloem sap contains very little amino nitrogen and aphids are good at assimilating most of it, honeydew is unlikely to be a source of nitrogen for ants. In addition to being a fuel for foraging, honeydew may also be stored and used to tide ants over periods of adverse conditions. A good example of this is the storing of coccid honeydew by honey-pot ants (Gullan and Kosztarab 1997). There is no example of this involving ants and aphids, but it could exist.

The cost for ants is that they need to monopolize, collect, transport and pass honeydew to their nest mates, which involves morphological and behavioural adaptations. However, the biggest cost is likely to be that associated with being dependent on aphids for fuel for foraging. This is particularly so for obligate myrmecophily, involving one species of ant and aphid. Although the distribution of the aphid *Stomaphis quercus* on oak trees is limited to those that grow within the territories of the ant *Lasius fuliginosus* (Latreille) (Goidanich 1956), it is unknown whether the distribution and abundance of the ant is dependent on that of the aphid, but it is unlikely because *L. fuliginosus* attends a wide range of tree-dwelling aphids. The expectation, however, is that ants should rarely be dependent on a single aphid species because this would put them at great risk of extinction.

Aphids are soft bodied and have little defence against natural enemies other than avoidance. Therefore, it is likely that a major benefit of ant attendance for aphids is protection. In habitats where aphids are at particular risk of attack from natural enemies, a high incidence of ant attendance is predicted. Most aphids that are ant attended are gregarious. Clearly, this is advantageous for ants because it results in the sources of energy being concentrated in a few places rather than scattered throughout their territory. However, in being gregarious, aphids become more attractive to natural enemies, which could put an upper limit on the size of ant-attended aphid colonies.

The cost for aphids appears to be mainly one of producing large quantities of high-quality honeydew to attract ants. It is well established that facultatively attended aphids increase their rate of honeydew production when attended by ants (Nixon 1951). Therefore, if unattended aphids feed at an optimum rate for the assimilation of amino nitrogen, then a faster rate is likely to adversely affect their feeding efficiency and consequently their rate of growth. If aphids are obligatorily ant attended, then another cost for aphids is the effect this has on their distribution. Good examples of this are the oak aphid, *S. quercus*, and the thyme aphid, *Aphis serpylli*, both of which have markedly more restricted distributions than their host plants because they are dependent on ants being present in the habitat (Hopkins *et al.* 1998, Hopkins and Thacker 1999). An overview of some of the hypotheses that are based on cost–benefit trade-offs of mutualistic interactions is given in Table 5.1.

Table 5.1. *Hypotheses proposed to account for ant attendance when framed within a cost–benefit perspective. These hypotheses are not mutually exclusive and some might apply to specific species only*

Hypothesis	Reference
Production of high quality, attractive honeydew is costly for aphids	(Fischer and Shingleton 2001, Yao and Akimoto 2001, 2002)
Physiological costs, like reduced growth rates when attended, must be traded against the benefit of protection against natural enemies	(Stadler and Dixon 1998a, Stadler 2004)
Aphids that feed on woody plant structures cannot quickly withdraw their stylets and escape natural enemies, and are therefore more likely to require the protection afforded by ants	(Dixon 1998, Shingleton *et al.* 2005)
Competition costs between different aphids and other honeydew/nectar resources for mutualists	(Addicott 1978a, Cushman and Addicott 1989, Cushman and Whitham 1991, Sakata 1999, Offenberg 2001)
Aphids living in dense aggregations are a better source of honeydew than solitary feeding individuals and the investment/costs per capita in large colonies are smaller	(Hayamizu 1982, Katayama and Suzuki 2002)
Carbohydrate/protein ratios of liquid food controls the identity of ant associates and the quality of the ant-rendered services	(Davidson 1997)
Mobile insects are less dependent on the protection services provided by ants. For example, they invest in wings rather than high quality honeydew. Winged morphs, however, are less fecund (more costly) than wingless individuals	(Dixon 1958, Stadler 2002)
Species of aphid that are covered with wax wool or winged should be less dependent on protection by ants. They trade off investment in own defence against that to attract ants for defence services	(Bristow 1991)
Some species of aphids trade off the cost of investing in soldiers against that in protection provided by ants	(Stern *et al.* 1995, Stern and Foster 1996)
Feeding on plants growing in N-rich soils favours the production of high quality honeydew and ant attendance. This honeydew is thought to be more attractive and less costly to produce	(Bristow 1991, Breton and Addicott 1992b, Collins and Leather 2002)
For many species of aphids, achieving high growth rates – small predator–prey generation times (GTRs) – is more important than investing in defence. (As a consequence, selection for benefits of protection services might be relatively unimportant)	(Dixon and Kindlmann 1998)
For ants searching for food is risky and costly. The ability to monopolize and transport honeydew to the nest needs behavioural and morphological adaptations. This needs to be traded off against a more opportunistic foraging strategy	(Fellers 1987, Davidson 1998)

Sometimes these hypotheses are not as explicit as our list of hypotheses might suggest but they show the main ideas and directions of research on ant–aphid mutualisms.

Ant and aphid colonies are female-dominated societies showing high degrees of relatedness. As a consequence, the costs for individuals and their contributions to future reproductive success of a colony are difficult to measure experimentally. Many studies show that when there is direct contact with predators ant-attended colonies do better than unattended colonies. The critical question is whether it is more rewarding for a *clone* to invest in ant attendance and benefit from the associated services or in some other form of defence or increase in growth rates. A potential drawback of the first option is the cost to obligate myrmecophiles of maintaining their investment in protective mutualism even when predator pressure is low.

From an ant's perspective it is equally interesting to ask whether it is more rewarding to exploit an ephemeral source of energy opportunistically or develop more specific behavioural and morphological features to exploit this resource, and how this ultimately affects the cost–benefit balance of a colony. There is no species of ant known that is exclusively associated with a particular species of Homoptera, indicating that a large degree of specialization is not advantageous for the better exploitation of aphids.

To give a more explicit example of fitness costs and benefits we report on one of our own studies, assuming that there should be at least two important components in the environment of an aphid, which affect costs and benefits and ultimately its fitness: plant quality and ants. To demonstrate the relative importance of these bottom-up and top-down components different aphid species that show a range of association with ants in high and low quality environments were studied. In contrast to other studies, which investigate the effect of specific environmental conditions on aphid performance, the fitness of different species of aphids in a high quality and low quality environment were compared. To do this, small plants of tansy were planted in 2-litre pots in a mixture of sand, gravel and compost in equal volumes (= low quality treatment, LQ) or in compost only (= high quality treatment, HQ) in a greenhouse, and workers of *L. niger* were given access to the aphid colonies or excluded. In this way four different environments were created for the four species of aphids. The aphids used were *Metopeurum fuscoviride*, *Brachycaudus cardui*, *Aphis fabae* and *Macrosiphoniella tanacetaria*. Their mutualistic relationships with ants have been well studied. *Metopeurum fuscoviride* is an obligate myrmecophile, *B. cardui* and *A. fabae* are facultative myrmecophiles (±), with the latter somewhat less intensively attended than *B. cardui* (Fischer *et al.* 2001), and *M. tanacetaria* is not attended (−). All of them feed on the upper parts of tansy,

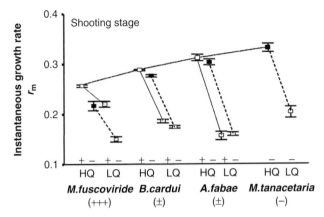

Fig. 5.2. Potential instantaneous growth rate, r_m for individual aphid species reared on high (HQ) and low (LQ) quality tansy when attended (+) and not attended (−) by *Lasius niger*. Similar relationships are observed at other plant growth stages. *Metopeurum fuscoviride*, obligate myrmecophile, +++; *B. cardui* and *A. fabae*, facultative myrmecophiles, ±; *M. tanacetaria*, unattended, −. (Modified after Stadler *et al.* 2002.)

which reduces the obscuring effects of different feeding sites, and allows one to focus on the effects of plant quality and ant attendance. Clearly, different species have different growth rates on the same host plant (Stadler *et al.* 2002) (Fig. 5.2). All species of aphids do better on high-quality plants. But the relative increase in the potential instantaneous growth rates, from LQ to HQ, is significantly higher in the unattended and facultative myrmecophiles, compared with the obligate myrmecophile, *M. fuscoviride*. In contrast, the fitness increase due to ant attendance was 23.5% in *M. fuscoviride*, 5.4% in *B. cardui*, but nil in *A. fabae*. The facultatively attended aphid species suffered larger costs on LQ-plants when attended by *L. niger* than the obligate myrmecophile, *M. fuscoviride*. This suggests that these aphids are adapted to seek out high-quality sites and suffer more if confined in growing aggregations by ants (= competition–colonization trade-off). In the best of all possible worlds the unattended *M. tanacetaria* did best (upper line in Fig. 5.2). These results show a complex mixture of costs and benefits for aphids when living on plants of different qualities and when entering an association with ants.

The above quickly leads to new questions. For example, how frequent are these particular environments and how does this affect the overall fitness of a clone? Moreover, could such heterogeneity lead to the shared exploitation of a common host plant? One way to approach the problem is to combine field data (from similar types of experiment to the above) and delayed life-history models (see Chapter 3) to account for the influence of past environmental conditions on current and future fitness and link the details of these life histories to population dynamics. This approach is especially useful for aphids because the impact of the maternal environment on offspring performance is well documented. For example, population size, N at time $t + 1$, can then be defined as:

$$N_{(t+1),q,a} = N_{t,q,a} R_{t,q,a} \frac{x_{t,q,a}}{1 + x_{t,q,a}} \tag{5.1}$$

$$x_{(t+1),q,a} = x_t \frac{M}{1 + N_{(t+1),q,a}} \tag{5.2}$$

The realized experimental population growth rate, $R_{t,q,a}$, is dependent on a point in time in the season, t, the quality of the host plant, q, which indicates nutrient availability within a patch (e.g. HQ, LQ) and the presence/absence of ants, a. M is a species-specific constant, which is adjusted to the field data. The results are not very sensitive to changes in M, which varied little within a treatment. Comparisons between treatments were made using optimized M-values.

The assumption that all patches are present in equal proportions (e.g. HQ + A, HQ − A, LQ + A, LQ − A), would result in an average performance of an aphid species in these environments. More telling, however, is to learn how these four aphid species will perform in habitats where the frequencies of patch qualities (plant quality, ant availability) vary. Some results are given in Table 5.2. When patch types are available in equal proportions (25%), or only the best patch (HQ + A) or the best two types of patches (HQ + A, HQ − A), counts of the obligatorily ant-attended *M. fuscoviride* are highest. If, however, only low-quality patches are available for aphids (LQ + A, LQ − A), either in equal proportions or only the LQ − A patch (100%), then the unattended aphid *M. tanacetaria* does best. Comparing the four species, neither of the facultative myrmecophiles ever attained the highest densities in these environments. The conclusion is that variation in environmental quality will produce multiple fitness optima and no single reproductive or mutualistic strategy will be the best under all conditions (Table 5.2). That is, different combinations of bottom-up (e.g. plant quality) and top-down (presence/absence of ants) effects lead to the simultaneous exploitation of a shared host plant by different species of aphids. The critical factor will be the relative frequencies of patches of different qualities/coverage at different times and the model indicates that we need to know more about the colonization potential of different species, in particular, the frequencies with which different aphid species are found on plants of different qualities (HQ, LQ, ± A) and subsequent probabilities of ant attendance.

Table 5.2. *Simulated abundance of four species of aphids assuming different relative frequencies of four types of patches (plants of high/low quality, HQ/LQ, and presence or absence of ants, +A/−A)*

	Relative frequency of patch types				
HQ + A	0.25	1	0.5	0	0
HQ − A	0.25	0	0.5	0	0
LQ + A	0.25	0	0	0.5	0
LQ − A	0.25	0	0	0.5	1
M. fuscoviride					
Sum *N*	11955.0	14834.0	15621.2	744.3	362.7
Final *N*	64.7	91.4	86.9	1.8	1.0
A. fabae					
Sum *N*	1724.6	1186.9	1951.6	413.6	213.2
Final *N*	3.9	2.9	3.7	1.7	0.9
B. cardui					
Sum *N*	1129.6	870.6	1114.0	435.1	201.7
Final *N*	4.9	3.6	4.7	2.1	1.0
M. tanacetaria					
Sum *N*	3467.9	1821.1	3681.6	1074.4	537.6
Final *N*	6.3	3.1	5.4	3.2	1.3

Sum *N* is the total number of offspring produced and Final *N* is the number of aphids at the end of the season.

Source: After Stadler (2004).

Aphids are clonal organisms and the ability to colonize new sites of high quality and escape competition or predation is important. There are many reports that show that ants prey on their homopteran partners (Pontin 1978, Offenberg 2001, Kay 2004), suggesting that ants may maintain a protein–carbohydrate resource balance or adjust the amount of honeydew produced according to their needs, that is maintain the monopoly over this resource. This happens even in obligate associations (Rosengren and Sundström 1991). However, whether this type of predation has a significant effect at the population level or over evolutionary time scales is unclear.

It is equally important to apply the cost–benefit perspective to ants. Ants do not simply collect honeydew but are also subject to predation risks and adjust their foraging activity accordingly (Carroll and Janzen 1973). For example, when food is offered to *Lasius pallitarsis* in patches where the risk of predation by *Formica subnuda* varies, *L. pallitarsis* spends less time foraging in the patches with *F. subnuda*, even though these patches contain high-quality resources (Nonacs and Dill 1991). Predator pressure on ants may also affect community structure. For example, in a manipulative field experiment, Gotelli

(1996) showed that ants use biotic cues associated with the presence of ant lions leading workers of different ant species to differentially allocate their foraging time in patches where these predators are present or absent. Forager abundance in pitfall traps was consistently lower in patches with ant lions and ants never foraged at high-quality tuna-fish baits in the presence of ant lions. Furthermore, the use of a high-quality patch depends on the magnitude of the difference in terms of growth between feeding in risky and safe patches: the greater the benefit of feeding in a risky patch, the more likely it will be exploited. Thus, ant workers are capable of evaluating risks and rewards and forage in a way that maximizes colony fitness (Nonacs and Dill 1990, Nonacs and Calabi 1992). This could imply that an aphid that is an obligate myrmecophile, which feeds on a plant in a habitat that is risky for ants, most likely will not be attended.

The relative and absolute benefits for ants when attending aphids or different species of aphids are less clear. From the point of view of ants the association is almost exclusively opportunistic. There is no case known where a particular ant species has obligate associations with a single aphid species. The 'price' of honeydew is dependent on the distance of the colony from the 'gasoline station' and the time required for collecting honeydew. Ants that are able to build their nest, or parts of it, close to such resources should have a selective advantage over those that are restricted in where they can build their nests. For example, the formation of subsidiary nests by *Formica rufa* leads to a network of intercommunicating nests with up to 200 meters between the main and peripheral nests (Gösswald 1941, Rosengren and Pamilo 1983). That is, polydomy could be an advantageous nest structure for species dependent on collecting large quantities of honeydew (Davidson 1997, 1998). For example, a colony of *Formica polyctena* may collect up to 240 kg of honeydew (fresh mass) during a season in southern Finland (Rosengren and Sundström 1991). This is mainly achieved by small peripheral nests, which may be temporarily established in the vicinity of honeydew resources.

5.3 The effects of ants on life-history characteristics and fitness

Studies on aphid–ant relationships often conclude that ant attendance has a positive effect on the fitness of aphids, resulting in larger colonies (El-Ziady and Kennedy 1956, Skinner and Whittaker 1981, Flatt and Weisser 2000), lower mortality rates (Way 1963) and less fungal infection due to the removal of sticky honeydew (Nixon 1951). However, such general statements do not account for the stunning fact that closely related aphid species have developed associations with ants ranging from close (obligate) through occasional (facultative) to avoidance (unattended) often in the same habitat. Only recently have the benefits which aphids are assumed to derive from ant attendance been questioned and critically examined. In particular, one needs to ask whether an increase in feeding and excretion actually benefits aphids, in particular those that are not closely associated with ants (see, for example, El-Ziady and Kennedy 1956).

Aphids do not simply tap into the phloem elements of plants and passively regulate the flow of plant sap through their bodies, which is sometimes collected by ants. They actively modify the composition of the sap in order to avoid dehydration (Fisher *et al.* 1984, Rhodes *et al.* 1996, 1997), which may also make it more attractive to ants and this could have significant metabolic costs. It is possible that osmoregulation is a preadaptation for forming an association with ants. For example, the aphid *Tuberculatus quercicola* incurs significant costs when attended by *Formica yessensis* as it produces smaller and less fecund adults than when unattended (Yao *et al.* 2000). Ant-attended aphids excrete smaller droplets of honeydew at a higher rate and the honeydew contains significantly higher concentrations of amino acids (Yao and Akimoto 2002), sucrose and trehalose (Yao and Akimoto 2001) than that of unattended aphids. Although the concentration of amino acids in honeydew is increased, possibly as a consequence of the effect of the increased flow of phloem sap on assimilation, its significance for ants needs to be established. It is suggested that in producing large quantities of honeydew aphids have less nitrogen for growth and reproduction (Stadler and Dixon 1998a). *Chaitophorus* spp. can also modify the sugar composition of their honeydew, and those that are more closely attended (*C. populialba, C. populeti*) are able to reduce the melezitose content when unattended (Fischer and Shingleton 2001). In contrast, *C. tremulae*, which is less often associated with ants, does not show this response. That is, the production of large quantities of honeydew, which is attractive to ants, is likely to incur fitness costs for aphids because they have to feed faster and increase the rate of converting simple to complex sugars. The sugar melezitose is thought to account for the attractiveness of aphids to ants and even preference hierarchies (Kiss 1981, Völkl *et al.* 1999, Woodring *et al.* 2004). The interpretation of this finding, however, has to be viewed in the context of the life-history characteristics of the partners, costs and benefits and population level effects on clonal organisms.

Clearly, population size can affect the relative magnitude of fitness costs and benefits. For example, individuals in small colonies of *A. craccivora* attended by *L. niger* excrete honeydew twice as fast as those that are unattended. In large colonies (more than 100 individuals), however, the excretion rates do not differ between attended and unattended individuals (Katayama and Suzuki 2002). Therefore, it is suggested that in large colonies of aphids individuals may benefit more from mutualism than it costs. However, further increase in population size is likely to increase competition for food and space and decrease the per capita protection benefit, making the net advantage of mutualism density dependent.

It is more difficult to demonstrate the fitness costs of ant attendance for obligate myrmecophiles than for facultative myrmecophiles because of the strong dependence of the obligate myrmecophiles on their partners. Facultative myrmecophiles are easier to manipulate. For example, contrary to the claim of El-Ziady and Kennedy (1956), *Aphis fabae cirsiiacanthoides*, feeding on *Cirsium arvense*, incurs significantly higher costs when attended by *Lasius niger*. It has a significantly lower mean relative growth rate, produces fewer offspring, takes longer to reach maturity and invests less in gonads (Stadler and Dixon 1998a). The facultative myrmecophile species feeding on tansy (*Tanacetum vulgare*) also have significantly lower potential growth rates when attended by *L. niger* (Stadler *et al.* 2002). Therefore, there is increasing evidence that aphids can change their physiology in response to ant attendance and this is costly. This new evidence needs to be added to the classical textbook paradigm of supposedly positive ant–aphid relationships.

5.4 Specialization and coevolution

Ant communities are thought to be organized by competitive interactions (Hölldobler and Wilson 1990, Porter and Savignano 1990, Vepsalainen and Savolainen 1990) and to a lesser extent by parasitism (Feener 1981) and predation (Gotelli 1996). Aggressive territorial species are unlikely to coexist, but submissive species are able to nest within the territories of dominant species. To escape competition many ant species forage opportunistically. That is, once resources become limited fugitive species are vulnerable to competition. Their success depends on escaping from resident competitors by arriving early in a new habitat. Species adapted to a fugitive way of life display a suite of traits, which includes early reproduction, the production of large numbers of relatively small offspring and dispersive morphs. This is the case for both ants and aphids. The ant species that are capable of quickly and effectively harvesting the energy of their homopteran partners (at least temporally) tend to be more numerous and dominant. As a consequence, the competition–colonization trade-off is one of the most common explanations and is the suggested prime mechanism for species diversity in ant communities and local co-occurrence (Stanton *et al.* 2002). In particular, the

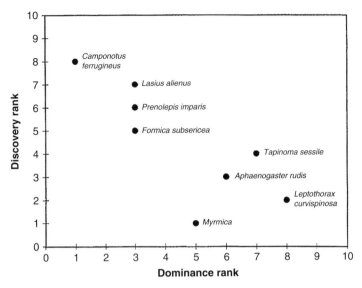

Fig. 5.3. The supposed trade-off in woodland ants between the ability to find and use resources before competitors (Discovery rank) and behavioural dominance over resources once located (Dominance rank). *Camponotus ferrugineus, Lasius alienus, Prenolepis imparis* and *Formica subsericea* are dominant, while the *Myrmica spp., Aphaenogaster rudis, Tapinoma sessile* and *Leptothorax curvispinosa* are subordinate. Aggressive dominants appear to be less able to find new resources quickly. Subordinate species, in contrast, quickly leave a resource in the presence of more aggressive species. (After Fellers 1987, Davidson 1998.)

competition-colonization trade-off is suggested for woodland ants (Fellers 1987) (Fig. 5.3).

The mechanisms underlying these patterns are numerous and include predation risk, behavioural and morphological traits, an ability to modify nest structure, habitat fragmentation or nutritional needs. For example, many ant species vary in their nutritional needs depending on the presence or absence of brood (Sudd and Sudd 1985, Portha *et al.* 2002). Species such as *L. niger* mobilize more foragers when feeding brood. Moreover, any excess of carbohydrate can be used to fuel greater activity and aggressiveness, maintenance of territories and competitive advantage over species that cannot effectively harvest honeydew (Davidson 1998).

Once ants evolved ways of collecting and processing honeydew it is likely that they were able to forage larger areas and become more abundant. Oster and Wilson (1978) distinguish between 'high tempo' and 'low tempo' ants. They assume that a positive correlation exists between behavioural tempo, colony size and polymorphism. Foraging over a large rather than a small area might impose a higher mortality risk, which may account for a higher turnover rate of workers. However, little attention was paid to the type of food collected and they predicted that homopteran-tending ants would be low tempo foragers. This is not in accord with the observed variation in foraging by ants associated with aphids and diet analyses, which also show a more complex picture with a high variability in the food types used (Blüthgen *et al.* 2003). The distance of the honeydew resource from the nest clearly affects the degree of attendance and smaller workers collect honeydew from colonies closer to the main nest (Bradley and Hinks 1968, Sudd 1983, McIver and Loomis 1993) (see also Section 6.4). As a consequence ant species that form satellite nests and shift to emerging honeydew resources might be more valuable to aphids than those that are less flexible in their nesting behaviour. Highly modified proventriculi are probably associated with large honeydew loads, and rapid ingestion of liquids is suggested to be of primary importance for the effective exploitation of honeydew by sap feeders (Davidson *et al.* 2004). As a consequence of this adaptation to this energy resource, higher species proliferation, for example in the genera *Camponotus* and *Formica*, contrasts with the species paucity (Bolton 1995) of genera such as *Dolichoderus*. Nevertheless, most North European ants are polyphagous, not specialized on any particular food and collect honeydew from a wide range of species (Stradling 1987). The conclusion is that the different degrees of mutualism between aphids and ants are necessarily based on short-term temporal advantages, behavioural and morphological adaptations and the spatial configuration of the partners because both groups of organisms are subjected to competition–colonization trade-offs.

Another difficulty in the evolution of a mutualistic association is the initial response of aphids to ants. It is often claimed that mutualism evolved out of a parasitic relationship, for example through the hosts' ability to terminate exploitation by a parasite (Johnstone and Bshary 2002) or individuals in populations with a mutualistic strategy doing better than those with an antagonistic strategy (Stachowicz 2001). However, a major evolutionary challenge for the development of mutualism is likely to have been the ability of aphids to survive encounters with ants (Doebeli and Knowlton 1998), especially as they are aggressive and ecologically dominant. Aphids, however, are a poor food resource (e.g. Toft 1995), because of their high sugar content, and are unlikely to be a preferred prey as long as alternative prey is available and ants are not recorded as specialist predators of aphids (but see *Lasius flavus*, Pontin 1978). In addition, recent studies on the aggressive invasive Argentine ant *Linepithema humile* revealed that surviving encounters with ants may not have been a serious evolutionary hurdle. Choe and Rust (2006) have shown that ants quickly associate a honeydew source with the chemical characteristics of the cuticle of the producer, in their study the brown soft scale *Coccus hesperidum*. In an elegant experiment they transferred the chemical cue in the cuticle of the scale to the cuticle of adult fruit flies and showed that the predation of treated flies by naïve ants decreased dramatically as they gained experience of tending brown soft scale. They suggest that this would enable ants to quickly adapt to new honeydew sources. Reducing aggression by trophallactic appeasement is a well-known phenomenon in the aggressive interactions between different species of ants (Bhatkar and Kloft 1977, Hölldobler and Wilson 1990). Sakata (1994) also reports that if an aphid produces honeydew in response to ant aggression and the honeydew is accepted by the ants, then predation of the aphid by the ant is much lower than when the aphid did not produce honeydew for the ants. Choe and Rust (2006) suggest that the ant/homopteran trophobiosis is mediated not only by a strong preference of ants, but also by a sophisticated learning process related to honeydew acquisition and associated cuticular chemical cues.

Mutualistic relationships also tend to be exploited by specialist predators and parasitoids or by cheaters (Bronstein 2001). For example, a number of coccinellid larvae are known to have evolved protective structures like waxy wool, chemical mimicry or inconspicuous movements. The wax-covered larvae of *Scymnus interruptus* and *S. nigrinus* survive attacks by *L. niger* and *F. polyctena* more often than larvae without wax (Völkl and Vohland 1996) and reach higher densities in ant-attended colonies, indicating ants probably afford the larvae protection against their natural enemies. Similarly, the wax-covered larvae of *S. sordidus, Hyperaspis congressus* (Bartlett 1961) and *Platynaspis luteorubra* Goeze (Völkl 1995) are reported to be protected. Larvae and adults of *Coccinella magnifica* use behavioural and chemical defence to avoid attacks by *F. rufa* (Sloggett *et al.* 1998, Sloggett and Majerus 2003). Aphid parasitoids, such as *Lysiphlebus cardui, L. hirticornis, Paralipsis eicoae* and *P. enervis*, have evolved chemical and behavioural mimicry, which enable them to exploit ant-attended aphid colonies without being attacked (Takada and Hashimoto 1985, Völkl 1992, Mackauer and Völkl 1993). It is, however, not clear how often such relationships have evolved, what the relative costs are for these specialist aphid predators and parasitoids or what kind of selection pressure they exert on aphid–ant relationships.

In addition, aphid–ant relationships may also be exploited by other species of aphids. As indicated above, an association with ants entails costs. Therefore, to obtain attendance benefits without paying the cost would be beneficial for an aphid. Although conceivable, there is no direct evidence that aphids are able to exploit established aphid–ant relationships, for example by benefiting from protection services on a shared host plant. The indirect evidence is that on the leaves of birch (*Betula pendula*) the co-occurrence of *Betualphis brevipilosa* with *Callipterinella calliptera*, which is ant attended, is more frequent than expected by chance, while co-occurrence with the unattended *Euceraphis betula* is random (Hajek and Dahlsten 1986). Therefore, not only interspecific competition for ants might act against the development of mutualistic relationships but also specialized natural enemies and opportunistic aphids.

When considering aphid–ant relationships one needs to appreciate how selection might act. Aphid colonies largely consist of clones; ants are socially organized with a high genetic relatedness and, as a consequence, selection in both taxa does not operate on individuals but on the colony or genet to which an individual belongs. Therefore, a useful approach is to attempt to understand how a clone, rather than an individual, should invest in defence and reproduction. An interesting case is aphid soldiers, which evolved in the closely related families Pemphigidae and Hormaphididae (Aoki 1978, Stern and Foster 1997). In a number of social aphid species it is known that on both their primary and secondary host plants the aphids may have different means of protection. The tropical aphid *Cerataphis fransseni*, for example, is attended by various species of ants on the secondary host but not on the primary host where this aphid produces galls (Stern *et al.* 1995). Similarly, *Pseudoregma sundanica* has two defence strategies. It is obligatorily attended by ants *and* produces sterile soldiers, and adjusts the level of investment in soldiers in response to ant attendance. Soldier production and ant attendance are negatively correlated (Shingleton and Foster 2000). This response is rapid and leads to a significant change in caste structure. However, only if soldier production is directly and inversely proportional to the incidence of ant attendance are the costs the same. Investment in ants and/or soldiers might be just two alternative ways of defence, equally costly, but with different pay-offs in different environments. Therefore, a major challenge is to identify costs and benefits at the level of a clone even if only the fitness of individuals can be studied. Preliminary molecular evidence on phylogenetic patterns in aphid–ant relationships indicates that ant tending is an evolutionarily labile trait, which was evolved and lost several times. For example, ant attendance in the genus *Chaitophorus* evolved at least five times; that is, it is relatively 'easy' for an aphid lineage to 'gain and lose tending' (Shingleton and Stern 2003). A similar conclusion was drawn from a comparative analysis of ant attendance of 112 species of aphids and 103 species of lycaenids (butterflies) from Europe, based on morphological and ecological traits, such as size, mobility, host specificity, feeding site, colony structure or host characteristics. Overall, the statistical analysis of these traits showed that in both groups of insects relationships with ants were only slightly (10%) associated with environmental and ecological traits; that is, the predictive power of these traits to explain ant attendance is poor. For example, for aphids, feeding on woody plant parts is significantly and positively associated with ant attendance, while the possession of wings in the adult stage is negatively associated with ant attendance. Lycaenids feeding on inflorescences or on nitrogen-rich Fabaceae host plants are significantly more likely to establish mutualistic relationships with ants. When this variation is broken down into different taxonomic levels, for aphids most of the variance in these traits is explained by ant attendance at the subfamily level and least at the species level, whereas for the lycaenids most is explained at a higher taxonomic level, such as a tribe (Stadler *et al.* 2003). This suggests that aphids are more flexible in their associations with ants, entering and leaving associations with ants whenever it is advantageous. Whatever the environmental conditions are, different species are able to adjust the degree of association with ants. The most likely reason for the difference is that honeydew is (at least partly) a waste product. Lycaenids, in contrast, have to invest in morphological structures such as nectar glands, the structure and function of which are less easy to modify. Therefore, once an association with ants is formed it is less likely to be discontinued. As said above, it is interesting that the relationships between ants and partners of ants are only to a small extent (10%) explained by coarse-grained morphological and ecological variables. This possibly indicates that the effects on net costs and benefits of life-history characteristics such as growth rates, developmental times, GTRs (see Section 6.1) and number of offspring or spatial characteristics of the habitat are more critical for the evolution of associations with ants. Currently, however, the published data on life-history traits are not sufficiently precise or systematic to address such a suggestion statistically.

Although the focus here is on mutualisms between ants and aphids it is interesting to point to similarities in the geographic evolution of virulence in host–parasite interactions. Theory predicts that the most virulent parasites should be found in the most productive environments and avirulent strains tend to dominate over virulent ones in less productive environments (Hochberg *et al.* 2000). Associated with this is the ease with which new hosts can be colonized and infected and the size of the susceptible host population (Ewald 1994). If parasites are easily transmitted from host to host (good colonizers) they tend to be more virulent than less mobile parasites. The latter are often more benign. With respect to mutualism between aphids and ants this suggests that in productive environments (e.g. those with a low predator pressure, both for aphids and ants) antagonistic and neutral relationships should prevail because little is gained from the services of a potential partner.

In contrast, when conditions are harsh, risky, or the environment is poor in available nutrients mutualism should be more likely to be found, because the specific services of the partner may be essential to exploit otherwise inaccessible resources. Similarly, if honeydew strengthens the competitive ability of ants then a benign behaviour of ants towards aphids should be expected. Otherwise, if there is a surplus of honeydew or nectar less specific/negative relationships between ants and aphids are likely and opportunism/antagonisms can be expected to dominate their relationships.

IN SUMMARY, aphids and ants provide resources, which are spatially distributed and vary in quantity and quality over time. Each mutualistic partner is thus subjected to significant spatio-temporal uncertainties with respect to resource availability, resource frequency, resource density, or potential payoffs when exploiting such a resource and it is reasonable to assume that the selection pressures of aphids on ants are as strong as that of ants on aphids. Two possible ways of decreasing the risk of being overexploited by their respective partners might be (1) to seek out new resources (colonize new habitats and initiate satellite colonies), which requires good discovery abilities and mobility, or (2) being competitively superior and monopolizing the available resources, for example a host plant or aphid colony, respectively. Both partners are therefore engaged and interconnected in competition–colonization trade-offs in a spatial setting in which density dependent costs and benefits and local top-down and bottom-up forces (Chapter 6) affect the outcome of the interactions, which can be positive or negative at the same time but at different places (meta-mutualism). However, the expectation is a mosaic of relationships, resulting in complex mutualism landscapes over which condition-dependent fluctuations in mutualism–antagonism evolve. Analysing the conditions that favour the local dominance of mutualism or antagonism across a heterogeneous landscape appears essential for understanding the origin and ongoing evolutionary dynamics in the range of mutualisms developed between ants and aphids.

6

Multitrophic-level interactions

6.1 Mutualism within a resource-tracking framework

6.1.1 Bottom-up and top-down forces

Interacting populations are subject to a host of biotic and abiotic factors, no matter whether the result of this interaction is obligate or facultative, positive or negative for the members of one or both populations. This is not new. Trophic relationships were included in the debate about the importance of resources (bottom-up) versus natural enemies (top-down) in determining population size and community structure. In addition, this food web perspective touches upon a century-old issue in ecology, which is the population regulation paradigm (Turchin 1999). This fundamental concept in ecology states that demographic density dependence is the key mechanism for population regulation. The simplest model of regulation is the logistic equation (Verhulst 1838) (for a full historical account see Kingsland 1995, Hixon *et al.* 2002). It is evident that natural populations do not grow unchecked, yet if and how population regulation occurs still remains an issue treated in special features in major ecological journals (Graham and Dayton 2002) and periodically resurfaces in skirmishes (Murray 1999, Turchin 1999). For example, in their highly influential paper Hairston *et al.* (1960) argued that in terrestrial communities, decomposers, producers and predators are resource limited 'in the classical density dependent fashion', while herbivores are controlled by predation; that is, they are unlikely to compete for resources. Not all agreed then and now (Power 1992, Dixon 2005). Turchin (1999) listed six points, which he proposed might provide a consensus of the views of population ecologists regarding population regulation (Table 6.1).

The different models introduced in Chapter 3 readily capture most of the points listed here. Yet, this view of populations is essentially that they are closed, which leaves an uneasy feeling as many important population features,

Table 6.1. *Issues of agreement amongst population ecologists according to Turchin (1999)*

(1) The realized per capita rate of population change ($r_t = \ln (N_t/N_{t-1})$) is subject to natural selection.

(2) r_t is affected by biotic (endogenous) and abiotic (exogenous) factors.

(3) The negative relationship between r_t and population density is a necessary (but not sufficient) condition for population regulation.

(4) Population dynamics are inherently nonlinear with a wide variety of functional relationships between r_t and population density.

(5) Rates of population change (dN/dt) may be affected by previous population densities (time lags).

(6) Rather than simply testing null hypotheses a more fruitful approach is to analyse time series in population dynamics and try to understand in what way exogenous *and* endogenous factors contribute to population change and regulation.

such as dispersal or habitat structure, with which an open population has to contend, are not included.

Given the difficulty of determining the negative feedback processes that keep populations within certain bounds it is not surprising that mutualisms played virtually no role in these discussions, which focused on predator–prey and host–parasitoid systems. Mutualistic interactions were simply not thought to modulate top-down and bottom-up forces or were perceived as destabilizing rather than stabilizing forces. However, the detailed information that is available for ants and their partners, at almost all levels of observation, provides an excellent opportunity to evaluate the different hypotheses in situations that include positive interactions. So, while these issues are not new, the discussion is no longer about whether bottom-up forces are more important than top-down processes or vice versa, or whether a population is regulated in a density dependent way or unregulated, but rather what controls the strength and relative importance of the influential variables under varying conditions, and what role mutualistic relationships play in structuring communities. Because ants and their herbivorous partners occur in intermediate trophic levels (see Fig. 1.1) mutualistic relationships between these groups of organisms are suitable for evaluating bottom-up and top-down aspects, density dependent processes, dispersal and community aspects in greater detail in order to understand the dynamic nature inherent in most relationships between ants and their partners.

So, what is the experimental evidence for bottom-up and top-down processes in interactions between ants and partners of ants and how do they affect the outcome of mutualistic interactions at the population level? Predominantly top-down forces are proposed for unattended *Aphis varians* on fireweed, *Epilobium angustifolium*, as natural enemies, such as coccinellids and syrphids, have a stronger (local) effect on aphid populations than water stress of fireweed (Morris *et al.* 2003). The magnitude of the effect of natural enemies remained the same over a period of three years. Similar top-down effects are reported by many authors, primarily for herbaceous plants and agricultural systems (Schmidt *et al.* 2003). In contrast, for tree-dwelling aphids, such as *Drepanosiphum platanoidis*, there is considerable evidence that natural enemies may not affect aphid numbers in a density dependent way (Dixon 2005). A strong argument for this is that the generation time of many natural enemies is much longer than that of their prey and therefore they are incapable of significantly reducing the prey populations at least during some part of the season (Kindlmann and Dixon 1999). In addition, the functional response, that is the number of consumed prey per unit time, must reach the upper asymptote. For example, Type II and Type III functional responses often lead to unstable predator–prey systems with natural enemies unable to regulate their host population (May 1973). In mutualistic systems functional responses might be more complex with respect to the ability of partners of ants to recruit ants and the subsequent effect this has on survival. Morales (2000a), for example, suggested a Type II recruitment response of ants of the genus *Formica* to increasing densities of *Publilia concava* on *Solidago altissima* and the benefit from ant tending for individual hoppers was highest in small colonies but decreased as aggregation size increased.

The ratio of the developmental time of the predator to the developmental time of the prey (GTR) might be a good proxy for the relative impact of natural enemies on herbivore populations (Dixon and Kindlmann 1998) which are unattended, but less so for herbivores that show some degree of association with ants. For example, the GTR of natural enemies and scale insects is about 1:1, while that of natural enemies and most aphids is substantially larger than 1 (Dixon 1998). Most likely, however, GTR ratios vary substantially over the season, depending on, for example, weather conditions, plant quality or the relative costs of the interaction for both partners. The rates of change in developmental times might be substantially larger for sap-feeding insects than for their natural enemies. For example, it is conceivable that after a period of optimal food availability while leaves are actively growing, rates of increase fall and developmental times rise for sap feeders (Fig. 6.1).

The increase in developmental time D should also occur some time before that of natural enemies because of declining plant quality or as a result of intra- and interspecific competition experienced with increasing population size of

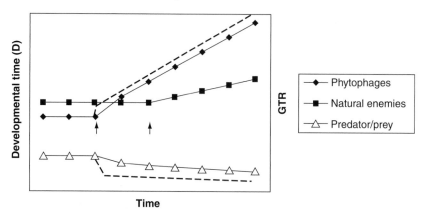

Fig. 6.1. Hypothetical developmental times of phytophagous insects (diamonds), their natural enemies (squares) and generation time ratio (GTR) of natural enemies/phytophagous insects (triangles) over a season. Dashed lines indicate prolonged developmental times due to ant attendance and a subsequent decline in GTR. Arrows give the point of time of increase in *D* in prey and predators, with the latter occurring later.

phytophagous insects; that is, well before predators experience a shortage of food (prey). As a result, GTR may be larger (above the 1:1 regulation line) early in the season and lower (below the 1:1 line) later in the season, which would make the relative impact of natural enemies on their prey a function of time (season). As a consequence, one would expect that with declining GTR protection by ants against natural enemies should become more important. However, ant attendance might prolong development because of the costs of producing high-quality honeydew and nectar; consequently GTR, with respect to ant attendance, might even further drop below the 1:1 line, increasing the risk for phytophages, especially when experiencing periods during which ants are not available. This leads to the interesting question: is the GTR of ant-attended aphids different from that of unattended aphids? Currently there is no evidence for this but it implies that seasonal shifts in bottom-up and top-down forces could have an impact on phytophagous insect populations.

Summarizing, the GTR principle might be a good first approximation to evaluate the potential role of natural enemies on population dynamics (for population regulation) and given that developmental rates are closely linked with fitness for different insect groups (Fig. 6.2) (e.g. the intrinsic rate of population increase, r_m) then this ratio might even be a good surrogate for the relative prevalence of regulation processes in natural populations. Currently, there are not enough data available to separate the relationships of ants and their partners showing different degrees of myrmecophily, but it is conceivable that per capita developmental rates of at least the facultative

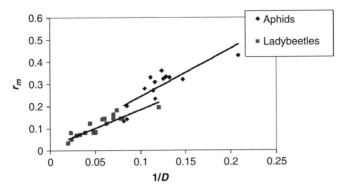

Fig. 6.2. Relationships between intrinsic rate of natural increase (r_m) and developmental rate ($1/D$), both measured at 20 °C, for various species of aphids (diamonds) and ladybird beetles (squares). In both cases the relationship is highly significant. (Data from Dixon 1998, 2000.)

myrmecophiles will decline when attended by ants and thus reduce fitness. This could lead to an optimal aggregation size and an optimal degree of attendance if there is a trade-off between maximizing the probability of tending (protection services) and minimizing competition for ant services and per capita costs of providing resources to attract ants.

The consumption of temporally limited resources by a growing population of phytophages will eventually lead to a shortage of this resource and a decline in the local carrying capacity. Most likely it will also lead to increasing competition between the members of a population and potentially promote dispersal. The extent to which different species experience competition on the same resource clearly differs, as does their ability to extract nutrients from the phloem sap of their host plant. These differences might depend on the ability of an aphid to induce changes in the host plant in a way that affects amino acid composition of the phloem and is positive for the aphid (Telang *et al.* 1999, Sandström *et al.* 2000) or its symbionts. As a consequence, the species-specific capacity to alter plant conditions may result in large differences in developmental and growth rates and competitive abilities over the season, ultimately affecting the community structure of sap-sucking insects (see Section 6.4).

There are many features of plant architecture and chemistry which should cascade from the bottom up through a trophic chain (Montllor 1991). Phytophagous insects, and sap-feeding insects in particular, are nitrogen limited because the phloem sap and especially xylem sap contains very little nitrogen (Mattson 1980). For example, fertilization of *Artemisia ludoviciana* with ammonium nitrate resulted in larger numbers of aphids (*Microsiphonella artemisia*) and membracids (*Publilia modesta*) with a concomitant increase in attending and patrolling ants (Strauss 1987). A positive response by sap-feeding insects to host fertilization has been documented for a variety of sap feeders like the green spruce aphid on Norway spruce (Straw and Green 2001), the grey pine aphid on Scots pine (Kainulainen *et al.* 1996), adelgids on hemlock (McClure 1980), different aphid species on tansy (Stadler *et al.* 2002) and, in particular, for economically important species in agricultural systems (Honek 1991, Awmack and Leather 2002). This has led to pondering whether host plant quality mediates aphid–ant mutualism (Breton and Addicott 1992b) and the concept of permissive host plants (Bristow 1991). The argument is that mutualisms between ants and their partners are resource driven, thus underestimating the active role partners of ants are able to play in controlling the extent of interactions with ants. However, analyses of honey-dew sugar composition showed that the same *Chaitophorus* species feeding on different *Populus* species produced honeydew with a markedly different composition with respect to the proportions of mono-, di- and trisaccharides (Fischer and Shingleton 2001). In addition, *C. populialba* and *C. populeti* are capable of increasing the production of melezitose when attended by ants and reducing the production of trisaccharides when unattended.

6.1.2 *Plant chemical defence*

The trophic link between plants, sap feeders and ants might be particularly affected by secondary plant chemicals such as alkaloids or tannins. One of the best studied systems is that of ragwort *Senecio jacobaea*, the aphid *Aphis jacobaeae* and moth *Tyria jacobaeae*. Ragwort is a perennial and flowering is often delayed for several years because a minimum size has to be reached before it will flower (Prins *et al.* 1990). The moth can locally defoliate ragwort, especially those plants with low pyrrolizidine alkaloid content. Ants may attend the facultative myrmecophile *A. jacobaea* and prey on the moth larvae feeding on these plants. Two scenarios are conceivable: (1) If aphids or ants select plants with a low pyrrolizidine alkaloid content then these plants should have a higher fitness in years when there are many caterpillars of *T. jacobaeae*, because undefoliated plants produce more seeds than those defoliated; (2) in years when there are few caterpillars those plants that are infested by aphids might have a lower seed output compared with uninfested plants (Vrieling *et al.* 1991). Thus, genetic variation in alkaloid content of *S. jacobaea* is thought to be maintained through interactions between the aphid–ant mutualism and the main leaf feeder of this plant. It is known that plants with aphids have lower pyrrolizidine alkaloid concentrations than plants without aphids. Unfortunately, there are no data on the magnitude of the fitness reduction due to aphid infestation or how ants respond to honeydew that contains high concentrations of secondary compounds. Therefore, it is unclear whether it is the aphid or the ant that selects plants of low alkaloid content.

In two similar studies the passage of quinolizidine alkaloid through the trophic food web was quantified using *Macrosiphum albifrons* and *Aphis genistae* on different host plants (*M. albifrons*: *Lupinus polyphyllus*, *L. albus*, *L. angustifolius*, *A. genistae*: *Petteria ramentacea*, *Sophora davidii*, *Spartium junceum*, *Genista tinctoria*). Quinolizidine alkaloids are characteristic secondary plant compounds in many *Fabaceae* and these substances are transported throughout the plants' vascular systems and eventually stored in vacuoles. These compounds are used by the plants to transport and store nitrogen and in the chemical defence against herbivores and pathogens (Hol and Van Veen 2002). *Macrosiphum albifrons* and *A. genistae* sequester quinolizidine alkaloids, reaching 4 mg/g fresh weight in *A. genistae* and 1.8 mg/g in *M. albifrons* (Wink and Witte 1991). It is suggested that these aphids are able to exploit the chemical defence compounds of plants for their own defence against natural enemies such as other insects or birds. For example, after feeding on *M. albifrons* the carabid beetle *Carabus problematicus* is paralysed for nearly 48 hours (Wink and Römer 1986) and the mortality rate of *C. septempunctata* larvae is 100% after five days when fed on alkaloid rich *M. albifrons* but only 20% when fed on alkaloid free *Acyrthosiphum pisum* (Gruppe and Römer 1988). *Aphis genistae* and *A. cytisorum* are facultatively associated with ants and the honeydew collected by *L. niger* workers from colonies of *A. cytisorum* contain on average 45 µg/g fresh weight of quinilizidine alkaloids (Szentesi and Wink 1991). While the ability to store these compounds is clearly beneficial for the aphids in their defence against their enemies it is not clear whether these compounds are also used by ants for chemical defence. Similar trophic relationships occur in the interaction between golden rain *Laburnum gyroides*, the aphid *A. cytisorum* and the attending ant species *Lasius niger*, *Formica rufibarbis* and *F. cunicularia*. The alkaloid content of the aphid is 182–1012 µg/g fresh mass and that of the ants 45 µg/g (Szentesi and Wink 1991). The alkaloid content of leaves shows a clear seasonal decline and is variable in different plant organs. Reproductive organs have the highest concentrations throughout flowering. This means that the degree of chemical protection of plants and aphids sequestering these compounds shows considerable spatial and temporal variability.

Thus, an important conclusion is that understanding the effects of herbivory on plant fitness and trophic interactions requires an understanding of the temporal pattern of resistance and measuring resistance at the 'wrong time' during a growing season could obscure interpretation. Even so it is unknown to what extent these secondary plant compounds affect the strength of the associations between ants and their partners. The protection by ants against natural enemies might be just one and probably not the best option available to herbivorous insects. It shows, however, that the role of secondary plant compounds reaches far up the trophic food chain and the ability to manipulate and tolerate the chemical defence of plants most likely is an important feature of the mutualistic relationships between phytophagous insects and ants. In the case of *S. jacobeae* increase in nutrient supply introduces further variation into this system because ragwort quickly adapts defence levels to nutrient availability by allocating more resources to growth (Hol *et al.* 2003). Increased nutrient availability, for example, is associated with a significant decrease in concentration of pyrrolizidine alkaloid in roots and shoots and an increase in biomass. This combination of reduced secondary metabolites and increased growth is likely to affect the associated food web.

6.1.3 Host plant heterogeneity

Plants are mainly supplied with nutrients in nature by detrivores, which recycle nutrients in dead matter. Soil animal mediated indirect modifications of plant performance include (1) changes in nutrient mineralization, (2) changes in soil structure, (3) grazing on mycorrhizal fungi and plant pathogens, (4) dispersal of plant-growth-stimulating micro-organisms and (5) hormone-like effects, to name a few (Scheu *et al.* 1999). These indirect effects of the soil biota ultimately act on plant performance and on the above-ground biota (Wardle 2002). However, these effects can be either negative or positive. For example, in microcosm experiments with individual plants, the aphids *Sitobion avenae* and *Myzus persicae* are differentially affected by different groups of soil organisms. Protozoa and Collembola significantly increase aphid performance but earthworms have no effect (Fig. 6.3) (Scheu *et al.* 1999, Bonkowski *et al.* 2001).

The presence of protozoa in the soil significantly increased total aphid biomass (2.6 fold) relative to the control and the earthworm treatment. In addition, the average number of nymphs per plant was significantly higher in the protozoan treatment. Presence of soil animals, in particular protozoa, resulted in an increase in the biomass and total nitrogen content of barley. It is likely that the higher nitrogen availability in the plants positively affected aphid performance. In a similar experiment Scheu *et al.* (1999) demonstrated that soil invertebrates affect above-ground herbivore performance. Earthworms and fungal-feeding Collembola significantly affected the performance of the aphid (*Myzus persicae*) on a grass (*Poa annua*) and a legume (*Trifolium repens*) in different ways. Aphid reproduction on *T. repens* was reduced in the presence

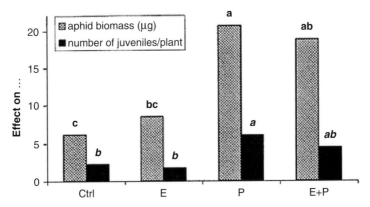

Fig. 6.3. Effects of protozoa and earthworms on the biomass of aphids (*Sitobion avenae*) and numbers of juveniles produced when reared on potted barley (*Tritium aestivum*) for two weeks in a constant environment. No soil organisms (Ctrl), earthworms (E), protozoa (P), earthworms and protozoa (E + P) added. Means of bars with the same letter do not differ significantly (ANOVA, $P > 0.05$, $n = 12$). (After Bonkowski *et al.* 2001.)

of Collembola by 45% but increased on *P. annua* by a factor of 3. The effects of earthworms were less consistent than that of Collembola. It is concluded that Collembola decrease reproduction on more palatable hosts, like *T. repens*, but increase it on less palatable hosts, like *P. annua*. This indicates that indirect effects of the soil fauna, such as the grazing of bacteria by protozoa and fungi by Collembola, might have a pronounced effect on nutrient availability, plant growth and aphid performance and above-ground community structure (Poveda *et al.* 2005). The soil fauna, therefore, may be more important than previously realized in determining above-ground heterogeneity in herbivore performance and community structure. It is, however, not clear whether soil biota can significantly affect mutualisms between partners of ants and ants, but it is conceivable.

The mosaic nature of host plant quality is especially pronounced in trees where different leaves and branches are exposed to different environmental conditions, such as exposure to solar radiation, water stress or attack by phytophagous insects and fungi that induces local defence responses in trees. An ability to track spatio-temporal changes in resource quality is important for maintaining high growth rates. The trade-off is between high mobility necessary for finding high-quality resources and a more sedentary and aggregated lifestyle necessary to meet the ants' demand for a continuous flow of honeydew. Mosaics of hosts may range from extremely vigorous to extremely stressed. Amongst the many hypotheses proposed to account for the interactions between plants and herbivores (Price 1997) there are two very different bottom-up hypotheses, which might account for differences in the abundance and performance of insect herbivores. The plant stress hypothesis (White 1978, 1984, 2004) suggests that host plants encountering physiological stress are more susceptible to insect herbivores than vigorously growing plants, because they are nutritionally more favourable and less defended due to constraints on their ability to mobilize secondary compounds. In contrast, the plant vigour hypothesis (Price 1991) suggests that vigorously growing plants are a more acceptable food for herbivores, because they are rich in nitrogen, which positively affects otherwise nitrogen-limited organisms. Unattended sap feeders, which are able to move between different parts of their host plant, should seek out the most nutritious parts. Tree-dwelling aphids, however, live in more heterogeneous environments. *Euceraphis betulae*, for example, may become significantly more abundant on stressed branches of birch (*Betula pendula*) late in the season when symptoms of water stress became apparent (Johnson *et al.* 2003b). *Euceraphis betulae* is also positively affected by a fungal pathogen (*Marssonia betulae*) of *B. pendula* and more aphids settle and achieve higher growth rates on infected leaves (Johnson *et al.* 2003a). *Drepanosiphum plata-noidis* shows a high incidence of movement between the leaves of sycamore (Dixon and McKay 1970) and typically feeds throughout the canopy early in the season, then moves to the lower canopy in summer, mainly to avoid heat stress, and subsequently occupies the whole canopy in autumn (Dixon 2005). Seeking out high-quality microsites in such heterogeneous environments requires a high degree of mobility with repeated periods when no sap is ingested or honeydew produced. This aphid has two peaks of abundance, one in spring and the other in autumn, and is in reproductive diapause during summer.

However, plant quality is only one of several factors that determine the quality of a feeding site. Large, mature leaves on flexible petioles brush against one another when it is windy, and dislodge the aphids. Sycamore leaves that have another leaf immediately below them are much less suitable for *D. platanoidis* than isolated leaves because in windy conditions the brushing action of the lower leaf dislodges many aphids, especially those that are unable to shelter along the main veins (Dixon 2005). Thus high-quality resources might be far less abundant than a green canopy might suggest. Most likely such chemically and physically heterogeneous environments and associated behaviour (high mobility) constrains the development of a close association with ants. Nevertheless, the different aphid species feeding on the same plant respond differently to these constraints. For example, of the 17 species of aphids feeding on birch in Central Europe, 3 are obligate myrmecophiles, 5 are facultative myrmecophiles and 9 are not associated with ants (Table 6.2).

Table 6.2. *Aphids on birch and their association with ants*

Species	Feeding site	Associated with ants
Euceraphis betulae	leaves	−
E. punctipennis	leaves	−
Monaphis antennata	petiole/leaves	−
Symydobius oblongus	twigs	+++
Glyphina betulae	leaves and young shoots	++
Callaphis betulicola	young leaves	−
C. flava	leaves	−
Callipterinella calliptera	leaves	±
C. tuberculata	leaves	±
C. minutissima	leaves	−
Betulaphis brevipilosa	leaves	−
B. quadrituberculata	leaves	−
Clethrobius comes	twigs	±
Stomaphis quercus	trunk, branches and twigs	+++
S. radicicola	trunk, branches and twigs	+++
Hammamelistes betulinus	leaves	±
Hormaphis betulae	leaves	−

−, unattended, ±, facultative, +++ obligate myrmecophiles.
Source: Kunkel *et al.* (1985), Blackman and Eastop (1994).

Although there is evidence that natural enemies can locally and temporally reduce the numbers of herbivorous insects, Hairston and co-workers' (1960) sweeping claim that phytophagous insects are top-down regulated needs qualifying. The notion that 'the world is green' implies that all foliage is of the same quality and there is little need to evolve exploitation strategies for competing for limited resources. However, for many sap-feeding insects bottom-up forces are generally stronger than top-down effects (Denno *et al.* 2000, 2003, Dixon 2005). Thus, the protection afforded by ants is likely to be less important for several groups of insect partners of ants than is generally assumed. For aphids, in particular, the ability to seek out resources of high quality, and maintain high rates of reproduction in combination with short developmental times is a defence strategy that enables them to outperform the functional and numerical responses of natural enemies. It stresses the importance of adaptive life-history features in a population and community context in determining the strength of top-down and bottom-up forces and on the location of a particular relationship between ants and their insect partner along the mutualism–antagonism continuum. It also indicates that models that incorporate life-history attributes are more likely to result in a better appreciation of regulatory processes. We concur with Price (2002) who says:

'any claim that top-down impact is stronger than bottom-up influences is necessarily couched in a narrow sense of biomass or numbers reduction'. An understanding of the role of these forces in regulating mutualistic systems depends on the relative importance of resource cascades up and down trophic levels, and measuring and understanding how organisms adapt to resource heterogeneity at all trophic levels in *open* systems. Mutualists that form distinct local colonies like ants and most sap feeders may show the character-istics of metapopulations but because they are positioned midway on the trophic ladder they are necessarily influenced by top-down and bottom-up effects.

6.2 Population effects

6.2.1 Seasonal dynamics and density dependence

Density dependent mortality and reproduction are not only central to the regulation of single species populations but may also play an important role in mutualistic interactions. The extent to which each partner invests in mutu-alism should depend on the density of the mutualist or, more precisely, the level needed to maximize its own fitness. This investment is likely to have a density dependent component. Whether this will result in stable or unstable population dynamics probably depends on habitat structure, the precision and time lags with which resources can be allocated to or removed from mutualists. Density dependent benefits for mutualists are known for a number of relation-ships between ants and aphids, ants and membracids and ants and lycaenids. For example, nymphs in large colonies of the membracid *Publilia modesta* benefited more from ant attendance (*Formica altipetens*) than when in small colonies (Cushman and Whitham 1989) (Fig. 6.4).

There are pronounced seasonal fluctuations in population size and differ-ences in the size of attended and unattended colonies occurred only in 1985 and 1987, but not 1986. Interestingly, only nymphs, but not adult membracids, were positively affected by ant attendance (Fig. 6.5). It is suggested that the age-specific benefit of ant attendance is a result of the nymphs being more susceptible to predation by salticid spiders, whereas the adults are more agile and better protected by their heavily sclerotized exoskeleton.

Research on the population dynamics of ant–aphid and ant–membracid systems indicates that aphids and membracids compete intraspecifically for access to ants (Wood 1982, Breton and Addicott 1992a, Morales 2000b). As the populations grow, the ratio of ants to membracids decreases, resulting in more membracids being vulnerable to predators in large than in small

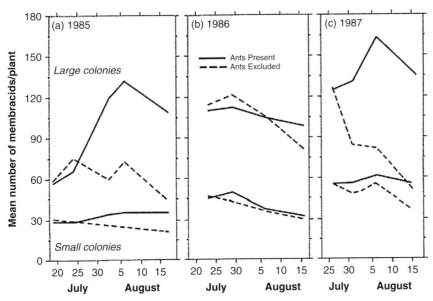

Fig. 6.4. Trends over a season in the population size of small (lower graphs) and large (upper graphs) nymphal colonies of the membracid *Publilia modesta*, in Arizona from 1985 to 1987. Solid lines represent the means for control plants (with attending ants), dashed lines those for experimental plants (ants excluded). (After Cushman and Whitham 1989.)

aggregations (Billick and Tonkel 2003). Therefore, colony size affects the outcome of mutualistic interactions and nymphs of membracids are more exposed to competition for mutualists than adults because of the density dependent response of ants. Multispecies relationships between ants and aphids also showed highly complex and density dependent benefits. For example, four species of aphids feed on fireweed (*Epilobium angustifolium*) in the Rocky Mountains and three of them are attended by 10 species of ants (Addicott 1979). The unattended aphid *Macrosiphum valerianae* was negatively affected by ants, *Aphis varians* and *A. helianti* were positively affected while *A. salicariae* was unaffected. For *A. varians* and *A. helianthi* the benefit of attendance was much larger for small than for large colonies in which aphid population size tended to decrease (Addicott 1979). Different species of ants had different effects. Thus, even on herbaceous hosts several strategies, ranging from antagonistic to mutualistic, are realized and species coexist. Newly established populations, consisting of only a few aphids, are more likely to become extinct than older populations that are ant attended. Given that the magnitude of the ant effect on aphids is large at densities or times when there are few aphids, and small or even negative when extinction is less likely, this could act to stabilize aphid abundance. This suggests that the numbers of

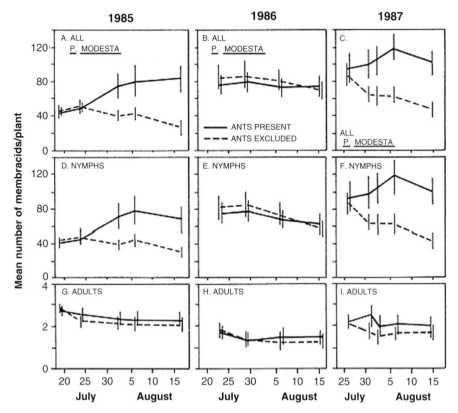

Fig. 6.5. Trends in population size of the membracid *Publilia modesta* on plants in Arizona during the growing seasons, 1985–1987. Solid lines represent the means for control plants (with attending ants), dashed lines for experimental plants (ants excluded). Means ± 1 SE. (After Cushman and Whitham 1989.)

ant-attended aphids should fluctuate less than that of unattended aphids. This idea is discussed further in the next chapter.

Density dependent effects on the outcome of interspecific mutualistic interactions are also recorded for ants preying on their attended partners. For example, *Lasius niger* collects honeydew from *Lachnus tropicallis* and *Myzocallis kuricola* on chestnut (*Castanea crenata*). However, this ant also attacks and eats the aphids depending on their abundance, but in an asymmetric way (Sakata 1995). *Lachnus tropicallis* appears to be the preferred honeydew source and workers attend the colonies in a density dependent manner. When the aphids are abundant *L. niger* attacks both aphid species, but more so the less preferred *M. kuricola*. It is suggested that predation is a function of the number of attended aphids per worker and it is the aphids that produce little honeydew that are attacked. Given that the level of mutualism has a density dependent component, one has to ask why this should be so and what are the temporal or spatial components of this predatory activity? The eating of aphids by ants has been reported on many occasions (Pontin 1978, Andersen 1991, Rosengren and Sundström 1991). However, it is unlikely that aphids are a preferred prey for ants and contrary to the suggestion of Sakata (1994) the reason for reducing the numbers of the partners is unlikely to be density per se or a tendency not to attack those species of aphid that provide large amounts of honeydew. Initially ants are likely to monopolize an aphid colony in order to prevent it from being taken over by a competing ant colony or another species of ant. This may initially facilitate the eating of aphids. But with further population increase ants may be unable to handle ever-increasing numbers of aphids, especially if their need for honeydew declines during the course of a season (Sudd and Sudd 1985). Therefore, a more useful approach to studying density dependent mutualisms is to quantify the costs and benefits associated with tending and eating aphids, and the seasonal changes in the strength of these associations, for example trade-offs between maximizing the probability of ant attendance and minimizing competition for ants when tended.

IN CONCLUSION, there is experimental evidence that density dependent processes commonly occur and possibly drive many of the associations between ants and their partners. Although density dependent processes are typically difficult to detect, or absent at times, because of many confounding variables, such as temperature or host plant quality, it is likely that they not only affect mortality and reproduction but also the relative benefits of partners in mutualistic associations. For example, the strength of density dependence in 32 independent populations of *Aphis nerii* was strongly correlated with population growth rates at low densities ($R_{initial}$) (Fig. 6.6) (Agrawal *et al.* 2004). That is, it is the aphid populations with high population growth rates at low density that experience the strongest density dependent regulation.

This is a robust result occurring in experiments performed at different times and different localities, irrespective of whether density dependence was affected by abiotic conditions or host quality. It is unclear, however, how density dependence varies in facultative and obligate myrmecophiles. Assuming logistic growth of populations implies that the strength of density dependence equals $-(r/K)$, and depends on how the growth rate, r, and the carrying capacity, K, scale. Three scenarios are possible. If K increases in the presence of ants then aphids with a greater r (e.g. obligate myrmecophiles) will have steeper negative slopes for density dependence compared with those that do not have a close association with ants (Fig. 6.7a); that is they can be expected to be relatively more affected by density dependent processes.

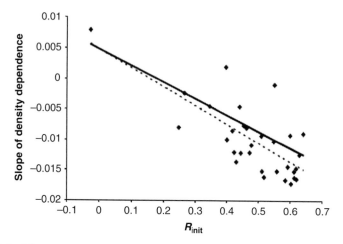

Fig. 6.6. The strength of density dependence across 32 independent experiments is predicted by the population growth rate (R_{init}) from each experiment. Each data point represents the slope and Y intercept from a regression of per capita growth rates on initial density. For heuristic purposes, the corrected covariance is portrayed as a regression line (solid line), calculated with the same intercept and variance in the independent variable as would be obtained from a simple regression with uncorrected data (dashed line). Accordingly, the covariance between X and Y is the only term that differs between the regression lines. (After Agrawal *et al.* 2004.)

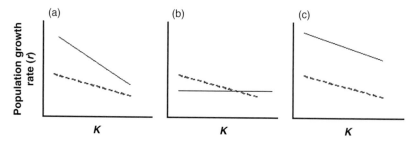

Fig. 6.7. A hypothetical depiction of density dependence in three populations which are affected (continuous line) or unaffected (dashed line) by ants. (a) With increasing K, density dependence is more pronounced in myrmecophiles with high r values, (b) K increases, e.g. because of protection from natural enemies, but r is negatively affected by ants, (c) ants affect both r and K, so the slope ($-r/K$) does not change. K = carrying capacity.

If r is negatively affected by ants but K increases (e.g. facultative myrmecophiles), then these aphids are expected to be less affected by density dependence (Fig. 6.7b). If, however, the ants have a proportional affect on r and K (e.g. obligate myrmecophiles), then the slope of the density dependent response does not change (Fig. 6.7c) and density dependent effects are expected to be identical in attended and unattended populations. There is a substantial literature on evolution and population biology that considers possible negative correlations between the two key parameters, r and K ($r-K$ selection theory). However, although these parameters summarize much of the ecology and evolution, it is unlikely that this simple relationship explains the evolution of mutualisms in all its facets. Nevertheless, as the strength of density dependence in many insects is strongly correlated with population growth rate at low densities (R_{init}) (e.g. aphids, Fig. 6.6) this provides some indication of the potential role of density in these systems. Thus, to incorporate the effect of density dependence into the population dynamics of mutualists, the magnitude, causes (e.g. differential benefits of different ant species) and correlates of variation in the strength of density dependence need to be understood.

6.3 Dispersal

A large number of the factors identified in experimental and theoretical studies select for increased dispersal in insects. Amongst these are avoidance of competition (crowding) (Dixon *et al.* 1968), spatial heterogeneity (Amarasekare 1998b, Plantegenest and Kindlmann 1999, Muller-Landau *et al.* 2003), kinship (Hamilton and May 1977), presence of predators (Dixon and Agarwala 1999, Weisser *et al.* 1999), or a combination of these factors (Ives *et al.* 1993, Kunert and Weisser 2003). For example, tending by ants can decrease the proportion of offspring that develop into winged individuals (El-Ziady and Kennedy 1956, Johnson 1959). The production of alates primarily occurs at high densities so there should be no effect of tending when population size is low. Whatever the exact reasons for dispersal, seeking a new host entails costs and benefits. For example, on an uninfested host aphids do not have to compete with conspecifics, there are no specialized predators or parasitoids and host plant quality is likely to be high. The downside, however, is that they might not find a new host and die during dispersal. An additional disadvantage for myrmecophiles is that they not only need to find suitable host plants but also sites where ants are available (Stadler *et al.* 2001). Given that dispersal has a strong effect on fitness, the timing and who should leave a host should be under strong selection pressure.

In spite of the detailed understanding of the many factors that affect dispersal, it is unclear how ants affect the abundance of their partners in general. For example, there is almost no information on how ants affect the large-scale population dynamics of their partners. Ant–aphid relationships provide a useful system for studying the population consequences of these interactions. Two important aspects need to be considered: (1) attended aphids might benefit from ant attendance due to protection from natural enemies, thus producing larger colonies; (2) ant attendance can have negative effects on aphids, at least on facultative myrmecophiles, for example as a consequence of the metabolism associated with producing more high-quality honeydew (Fischer and Shingleton 2001; Yao and Akimoto 2001). This leads to lower population growth rates in facultatively attended aphid species (Stadler and Dixon 1998a; Yao *et al.* 2000; Stadler *et al.* 2002). Consequently, the expectation is that in ant-attended species dispersal is delayed and their year-to-year population fluctuations may be less variable than in unattended aphid species. How can these hypotheses be tested? Long-distance dispersal of aphids is monitored by a network of suction traps used to forecast the abundance of pest species, to assess the infestation risk of crops and for providing recommendations regarding the actions to be taken to reduce the impact of aphids on crops. Because the data sets are biased to economically important species care must be taken not to overinterpret the results. Nevertheless, some interesting observations do arise (Fig. 6.8).

Figure 6.8 shows the average week of peak dispersal of unattended and ant-attended agriculturally important species of aphid at seven sites in France, based on suction trap catches. The peak in dispersal of ant-attended species is 1.5 to 2.5 weeks later than that of unattended species at all sites except Montpelier. Ants, site and year significantly ($P < 0.01$) influenced the timing of dispersal. Interactions, except for ants × site and site × year, were not significant. The significant interactions between the factors ants and site

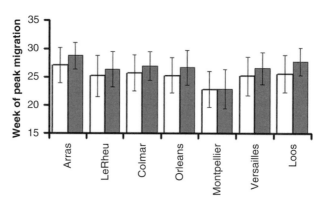

Fig. 6.8. Timing of peak migration of winged aphids in different suction trap locations for ant-attended and unattended species. Data are from seven different suction traps located in France and for a maximum of 22 years. Open bars indicate unattended aphids; closed bars, attended aphids. Vertical bars indicate SD. In total 35 species were identified in these catches. (After Kindlmann *et al.* 2007.)

and between site and year are a consequence of the north–south or coastal–continental gradients in the locations of the suction traps. It is likely that ant abundance or temperature dependent processes change along these gradients. However, the presence of these interactions does not affect the general conclusion. In addition, Slogget and Majerus (2000) purport to show that non-ant-attended compared with ant-attended aphids are scarce on trees from late July onwards. This suggests that mutualistic associations between aphids and ants affect large-scale population features such as long-distance dispersal and that the potential consequences associated with a delayed dispersal should also be considered when studying the costs and benefits in mutualistic associations. For example, a delay in dispersal of 1.5–2.5 weeks means that a complete generation is lost on a new host, compared with that achieved by conspecifics that are unattended. Currently, we do not know whether ants similarly affect their other insect partners, but it is likely.

Active dispersal by winged morphs is just one method of reaching another host plant. Circumstantial evidence suggests that aphids can use ants as vehicles to reach other host plants or plant organs. For example, Goidanich (1956) observed that the ant *Lasius fuliginosus* transports eggs of the obligate myrmecophile aphid *Stomaphis quercus* from its overwintering sites at the base of oak trees to suitable feeding sites in the canopy. More recently, Collins and Leather (2002) report *L. niger* transporting third and fourth instars of a facultative myrmecophile, the black willow aphid, *Pterocamma salicis*, to another host sapling, which was uncolonized. However, adult aphids can also easily disperse, and those with long legs are highly mobile. Thus, the importance of ant-mediated dispersal for myrmecophile aphids, relative to that achieved by winged morphs or by walking, is unclear. Amazingly ants seem to be able to transport aphids in their mandibles to plants of high quality, where the aphids attained higher growth rates. It is not clear, however, how ants recognize what are high-quality plants for aphids, but potential clues might be the chemical composition of honeydew or the restlessness of aphids. Given that aphids should always seek feeding sites of the highest quality, attended aphids should not be confined to specific feeding sites once attended. However, ant-borne dispersal appears to be a risky way to reach other host plants or plant organs, especially as aggressive ant species cannot be appeased due to the interrupted flow of honeydew. For long-distance dispersal, ants are an unsuitable means of transport, and given that fewer alates are produced by attended colonies, reducing the probability of reaching hosts of high quality (e.g. hosts at higher altitudes which develop leaves and shoots later in the season), attended colonies might experience significant disadvantages relative to unattended aphids. Long-distance dispersal in aphids is estimated to be hazardous, with approximately only 0.6% of the autumn migrants of the bird cherry-oat aphid *Rhopalosiphum padi* successfully colonizing the winter host (Ward *et al.* 1998). This suggests that if migration back to the winter host is delayed by ants, ant-attended host-alternating aphids should be at a selective disadvantage compared with those not attended.

In their models Dixon and co-workers (1993) and Poethke and Hovestadt (2002) make some interesting predictions about density- and patch-size-dependent dispersal rates. Even though their models were not derived with respect to ant attendance they can be easily modified and used to predict the consequences of myrmecophily.

The assumptions are that individuals will base their decision whether or not to migrate on patch capacity (patch size) and pre-dispersal population density. Dispersal will be beneficial if the fitness of the migrant is greater than that of a resident individual $F_M > F_R$. At very low population densities the expected fitness of individuals staying in the patch is greater than that of those that migrate. In this situation it will be more beneficial to stay (Fig. 6.9a). With increasing population size individuals should disperse at a rate d. Ants are likely to increase the population size at which myrmecophiles start to produce alates because they protect them against natural enemies. However, with increasing population size of the phytophages the negative effects of deteriorating host plant quality should increase quickly. Thus, with ever-increasing population size one would expect the bottom-up effects could become more severe than top-down effects, forcing the partners of ants to produce more and more migrants because F_R declines to very low values. This should lead to a large exodus of mutualists of ants. This model, however, does not account for

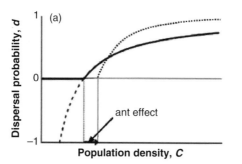
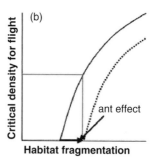

Fig. 6.9. (a) Dependence of the probability of dispersal on population size in myrmecophiles and non-myrmecophiles. (After Poethke and Hovestadt 2002.) (b) Incidence of migration in relation to habitat fragmentation. (After Dixon *et al.* 1993.) Larger population size and more fragmented habitats should lead to an increase in probability of dispersal and increase in the threshold population size for dispersal, respectively.

the effects of kin competition and consequences for inclusive fitness, which may affect dispersal probabilities. Thus this model may be more appropriate for membracids or lycaenids.

A more explicit consideration of habitat fragmentation and relatedness is included in the models of Dixon *et al.* (1993) and Plantegenest and Kindlmann (1999). For aphids it is suggested that if habitat fragmentation is low then migration occurs at low densities and probably more frequently compared with habitats with a high degree of fragmentation (Fig. 6.9b). From the perspective of a myrmecophilous partner ants are another component of habitat fragmentation. For example, for two species feeding on the same host, with one attended and the other unattended, the habitat should be more fragmented for the obligate myrmecophile than for the unattended herbivore. If the services of ants are essential, then once found in a patchy environment emigration from this patch is less likely. Therefore, as suggested before, the expectation is that ant attendance is likely to delay dispersal and to increase the critical density for dispersal, because not only new plant patches need to be found but also patches with ants.

Individuals in populations of clonal organisms, such as aphids, are likely to show a high degree of relatedness, which can be expected to have an effect on dispersal. Exploring the effect of habitat fragmentation on dispersal in clonal organisms is illuminating. Following the approach of Plantagenest and Kindlmann (1999) it is assumed that there are a large number of patches, all of which have the same properties. Local population dynamics is described by the Ricker equation (Ricker 1954), which separates population growth and dispersal in two time steps.

Population growth is:

$$N'_{t+1}(i,j) = N_t(i,j) \exp\left[\ln(\lambda)\left(1 - \frac{Npatch_{t(j)}}{K}\right)\right] \tag{6.1}$$

$N_t(i,j)$ is the number of individuals with strategy i in patch j at time t. N_{t+1} is the population size at the next time step, but before dispersal. λ is the finite rate of population increase. $Npatch_t(j)$ gives the total number of individuals in patch j, at time t ($Npatch_t(j) = \Sigma_i N_t(i,j)$), and K is the carrying capacity.

Dispersal is given by:

$$N_{t+1}(i,j) = N'_t(i,j)(1 - S_{t+1}(i)) + \frac{Ntot_{t+1}(i)(1 - m)S_{t+1}(i)}{Npatch} \tag{6.2}$$

The idea is that not all migrants succeed in locating a new host. Here, a fixed proportion of aphids is assumed to migrate in each time step. $N_t(i,j)$ is the number of individuals of clone i, in patch j, at time t that stay in the patch. $S_t(i)$ gives the proportion of individuals of clone i migrating at a particular time step t, and $Ntot_t(i)$ gives the total number of individuals of clone i ($Ntot_t(i) = \Sigma_i N'_t(i,j)$). The parameter m gives the proportion of individuals dying at each dispersal time step.

Next, evolutionarily stable strategy (ESS) and mutation rule are defined.

A mutant with a new strategy (e.g. earlier or later dispersal, different intensity of dispersal, exploitation of ants) will increase its frequency in the population at the end of the season if its relative rate of increase is larger than one:

$$\frac{Ntot_{Tfin}(mutant)}{Ntot_{Tfin}(resident)} \times (Npatch \times Ninit - 1) > 1. \qquad (6.3)$$

In this case the strategy of the resident is considered not to be evolutionarily stable. As a consequence, this strategy is eliminated and the mutant becomes the new resident.

Mutations affecting dispersal can be manifold. For example, dispersal can start earlier or later, or the intensity of dispersal can increase or decrease. Modification of the time and intensity of dispersal is achieved by selecting a variable from the series (S_t) at random, which determines the strength of the effect of the mutation (Plantegenest and Kindlmann 1999). The type of mutation is chosen at random but with equal probability. The model was run for 10 000 years with the initial values of the proportions of residents and mutants (i) migrating in each year chosen at random from the series (S_t).

Given that there is only a single foundress, the resulting high relatedness of the aphids in a colony leads to ever-increasing rates of dispersal at ESS, whose magnitude depends only on survival during dispersal (Fig. 6.10). With increasing survival (decreasing mortality) the incidence of dispersal at ESS quickly increases. That is, competition between members of the same clone is easily reduced by dispersal and the evolutionary benefit is large, if the cost of dispersal is low. Alternatively, if there is initially more than one foundress then the incidence of dispersal decreases. This is especially so if few survive dispersal (Fig. 6.10). That is, it is not advantageous for a clone to leave a host to find another one, because leaving only benefits the mutants, which prefer to stay on the plant. By leaving and reducing, the colony size benefits the other clones. Only if the success of travelling to another host is very high does it pay clones in a mixed genotype colony to migrate because this reduces competition both within and between genotypes.

Using a simple logistic equation means that competition between individuals increases with every newborn added to the colony (the per capita growth rate

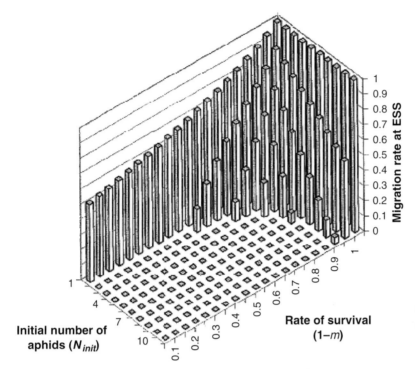

Fig. 6.10. Effect of different numbers of foundress of aphids (N_{init}) initiating a colony on the proportions migrating for different levels of survival during dispersal. More foundresses result in a decrease in the genetic relatedness of individuals in a colony. (After Plantegenest and Kindlmann 1999.)

dN/Ndt decreases linearly with increasing N). This is unreasonable because it leads immediately to dispersal and a reduction in intraspecific competition. In real plant–herbivore systems competition probably does not occur when few individuals use the same resource. As indicated in Section 3.3, introducing an Allee effect is useful for modelling situations in which initially the increase in numbers benefits each individual in a colony, then population size has a negative effect when competition occurs. That is, a minimum number of individuals is necessary before the per capita growth rate becomes positive. In the case of herbivores this could be because (1) more individuals reduce an individual's risk of falling victim to a predator and (2) a certain minimum number of individuals is needed to overcome the defence responses of the host plant. For example, the amount of saliva injected by a group of aphids might keep the sap flowing or improve its quality more than the saliva of one or a few aphids (Prado and Tjallingii 1997). Incorporating this into the Ricker equation (6.1) gives:

$$N_{t+1}(i,j) = N_t(i,j) \exp\left[\ln(\lambda)\left(1 - f(Npatch_{t(j)})\right)\right] \qquad (6.4)$$

In the previous simulation $f_1(N) = N/K$. An Allee effect can be incorporated into this equation by changing $f_1(N)$ to $f_2(N) = 4/K^3 N^3 - 1/K^2 N^2 - 2/KN$. There are many different ways of incorporating both positive and negative effects of density dependence on per capita growth rates (Allee effect) (Aviles 1999) but for illustration purposes we follow the phenomenological example of Plantegenest and Kindlmann (1999).

As expected the introduction of an Allee effect results in a delay in the onset of dispersal (Fig. 6.11). In this simplistic model the optimal time for dispersal is always when the population size is at $K/2$. If the ants increase the carrying capacity of the partner (Pierce and Young 1986, Neuhauser and Fargione 2004) as is the case for many myrmecophiles, then the optimal time for dispersal for aphids is further delayed and fewer migrants are produced (ant effect in Fig. 6.11). For example, the carrying capacity might be increased because there is less or no risk of predation in ant-attended colonies. That is, the period of co-operation in a colony of myrmecophiles is longer than for non-myrmecophiles. A reasonable explanation is that they have to co-operate in order to produce large quantities of honeydew to attract ants. Thus overall, an Allee effect might be more important for myrmecophiles than for non-myrmecophiles. There is little work on how ant attendance affects the population dynamics and migratory behaviour of the partners,

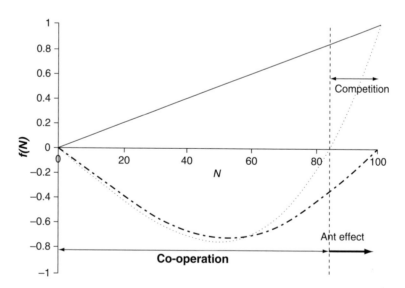

Fig. 6.11. Comparison of different functions incorporating no density dependent effect on per capita growth rates, $f_1(N)$, solid line; an Allee effect, $f_2(N)$, dotted line (K = 100); and an Allee effect with superimposed ant effect, $f_3(N)$, broken line (K = 120). (Modified after Plantegenest and Kindlmann 1999.)

but the results for aphid dispersal, described above, lend some support to the idea that dispersal of ant-attended species of aphid might be delayed (Fig. 6.8) (Kindlmann *et al.* 2007).

Habitat characteristics and migratory ability generate a trade-off between investment in reproduction and investment in lipid reserves for flight. Dixon *et al.* (1993) suggest that species living in non-fragmented habitats, like the canopies of trees, should always be winged and capable of flight and invest relatively little in lipid reserves in adult life. This allows them to quickly track resources without having to walk down a branch and back up a neighbouring branch. In contrast, species that live in highly fragmented habitats, like those that feed on herbaceous plants, need to stay for longer and the intervals between flights might exceed the lifetime of an individual. In this case it is advantageous to produce winged individuals only for long-distance dispersal, otherwise only unwinged individuals. This trade-off between allocating resources to gonads versus developing a flight apparatus seems to exist in aphids, with tree-dwelling species of aphid (living in unfragmented habitats) more likely to retain their wings and gonads and invest less in lipid reserves compared with those living on herbaceous plants (spatially unpredictable resources, fragmented habitats).

The widespread occurrence of flight polymorphism in insects strongly suggests that there are fitness costs associated with the ability to fly. For example, fitness trade-offs between flight capability and reproduction are reported for aphids (Walters and Dixon 1983, Dixon *et al.* 1993), grasshoppers (Ritchie *et al.* 1987), crickets (Roff 1984, Mole and Zera 1993, 1994, Zera and Mole 1994, Zera *et al.* 1994), planthoppers (Denno *et al.* 1989) and seed bugs (Solbreck 1986), with the flight-capable forms less fecund than the flightless morphs. The fitness penalties are either a reduced fecundity or delay in when they first reproduce, which in particular is a drawback for multivoltine species. Because of the internal resource-based trade-offs between flight capability and reproduction the maintenance costs of a flight apparatus should be minimized unless wings are needed by an organism for tracking changing resources several times during its lifetime. Thus, one would generally expect that migratory ability should be minimized in persistent habitats. Indeed, there is a large body of theory that predicts elevated levels of dispersal in ephemeral, patchy habitats (Southwood 1962, Roff 1986, 1990) with the important point that the length of the period of the persistence of a habitat needs to be measured relative to the length of the life of an individual or member of a clone. Many empirical studies support the hypothesis that the incidence of dispersal is inversely related to habitat persistence. For example, for 35 species of planthoppers inhabiting scrubby vegetation and grassland, there is a significant negative correlation between dispersal via winged morphs and persistence of their habitats, measured as the maximum number of generations attainable (Denno *et al.* 1991).

Besides habitat persistence habitat structure is also an important factor influencing the evolution of dispersal capability (Wagner and Liebherr 1992, Roff 1994). For example, wings may function to quickly navigate complex three-dimensional habitats, which are easy to reach by flying but not by walking. Relocation of feeding sites, following escape from natural enemies or when brushed off by leaves touching, may prove difficult for wingless individuals on trees. In contrast, the consequences of falling from a herbaceous plant, which produces many shoots, might be low because resources are easily relocated by walking (Gish and Inbar, 2006). Consequently, selection may favour the retention of flight capability in three-dimensional habitats like the canopy of trees, even if this type of habitat is persistent. There is evidence of this for one subfamily of aphids, Drepanosiphinae (Waloff 1983, Dixon 1984) and planthoppers (Denno 1994), which confirms that species that regularly produce unwinged forms less commonly inhabit trees than low growing herbaceous plants.

With respect to myrmecophily this means that species that invest in high mobility are less likely to be ant attended and confined to particular feeding sites than are apterous species. This is because individuals that frequently change their feeding sites, in order to track resources, are unlikely to be suitable partners for ants because of their aggressive response to moving objects. In addition, retaining wings allows adults to rapidly escape attacks by natural enemies. For example, in a study of 148 species of aphids most unattended species were on trees and bushes and few on grasses (Fig. 6.12). Most of the attended species are wingless and feed on woody plant parts, which are difficult to penetrate and to withdraw stylets from when disturbed by natural enemies (Shingleton *et al.* 2005). In these relatively rare cases it may be more advantageous to be associated with ants than to maintain flight capability.

IN CONCLUSION, Sections 6.1–6.3 indicate that bottom-up/top-down processes, density dependent/density independent processes and habitat characteristics via dispersal affect the outcome of the interactions between ants and their partners. These are scale-dependent issues and affect all populations of potential mutualists. A combination of theoretical and empirical studies that bridge the gap between small-scale studies and large-scale phenomena will provide further insight into the mechanisms that are likely to be scale and time dependent (e.g. species dispersal characteristics) and those that are scale independent (resource availability and quality).

Fig. 6.12. Percentage of aphids not attended (open columns), facultatively (dashed columns) and obligately (black columns) ant attended when feeding on trees/shrubs, herbs and grasses. (After Stadler and Dixon 1998b.)

6.4 Community effects

Local assemblages of interacting populations are connected in a myriad of ways via trophic food webs to outside influences (Polis *et al.* 1997). These interactions may occur between individuals at the same trophic level or between individuals of different trophic levels (bottom-up; top-down). If these relationships are a consequence of the physical interactions, such as between predators and prey, lycaenids and ants or plants and herbivores, then such interactions are termed direct (Holt and Lawton 1994). Additionally, interactions may be indirect; that is, individuals of one species do not physically interact with the individuals of another species (Wootton 1994). A prerequisite of indirect effects is that they require the presence of a third species to convey these effects. Indirect biotic effects are thus the exclusive property of multispecies communities. It is the nature of indirect interactions that they are more difficult to observe than direct pair-wise interactions, but there is now considerable evidence that they play an important role in determining community structure. Wootton (1994) defines five types of indirect interactions (Fig. 6.13).

Here this classification is adapted for the interactions in *local* species assemblages involving ants and their insect partners. Interspecific competition as depicted in Fig. 6.13a is a form of exploitative competition if two myrmecophiles (A, C) compete for the services of ants (B), which is well documented. Trophic cascades are indirect effects of species A on species C via its direct effects on species B. The mutualism between ants (A) and a herbivore partner (B) can reduce the growth or seed output of the host plant (Fig. 6.13b). Again

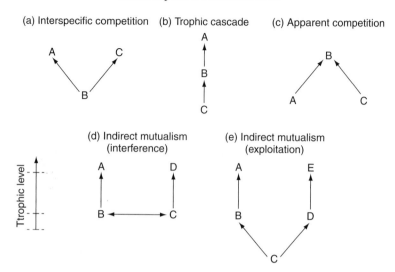

Fig. 6.13. Five types of indirect interactions observed in field studies. Arrows point to the individual of a particular species (A–D) that experiences an interaction effect (energy flow or service). These effects can be negative (e.g. competition, predation) or positive (mutualisms). Interference means that certain individuals in a population acquire an adequate supply of a limited resource (e.g. protection against natural enemies) and in so doing also benefit individuals of another species. Indirect mutualisms involving exploitation involve a negative effect of species C on species B and D, which might release species A and E from predation. (After Wootton 1994.)

there are many examples of these indirect effects. Apparent competition arises when two prey species share a common predator (Holt 1977) (Fig. 6.13c). Apparent competition may be highly asymmetrical, if one 'prey' species (e.g. A) is a facultative mutualist of ants. Then the negative effect of a predator might be larger on species C, at least as long as ants are interested in the resources provided by A. Indirect mutualism is an indirect positive effect of one species on another (Fig. 6.13d). An example might be the interaction between a myrmecophile species of aphid (A) and a plant with extrafloral nectaries (D). Some dominant ants might be more closely associated with homopterans producing honeydew, while subdominant ants might preferentially collect nectar from nectaries (Blüthgen and Fiedler 2004a, b). If the damage caused by the homopterans is low the constant provision of sugar resources might lead to a positive indirect interaction between Homoptera and nectaries or food body bearing plants. However, we know of no study of such a positive indirect interaction in plant–homopteran–ant systems. Exploitative indirect mutualisms (Fig. 6.13e) are similar to the situation depicted in Fig. 6.13a connected with an additional trophic level. For example, coccids (B) and lycaenids (D) might exploit the same host plant (C) and be attended by different non-aggressive species of ant (A, E), and benefit from co-occurrence on the same plant because both species of ant ward off general natural enemies. However, there are no examples of this, most likely because the relative benefits are difficult to quantify.

The above description is one way of imposing order on an indefinitely large variety of indirect interactions in communities composed of ants and their insect partners. However, this static view is less suitable when trying to understand the dynamics of these associations and especially the transitions from antagonism to mutualism and vice versa, which was discussed previously. In order to better understand the dynamics and constraints in these indirect community effects we explore a simple model by Abrams et al. (1998) who studied the indirect interactions between two prey species and a shared predator, when all the species undergo population cycles. Again, we present the model in the context of ants and their insect partners, which might be potential prey of the ants or mutualists. Abrams and co-workers start with the theta-logistic equation, which is a simple 'add-on' of the logistic equation with the parameter theta (Θ) determining the force of density dependence. For simplicity, assuming that $\Theta_i = 1$, they arrive at the following equations:

$$dN_i/dt = N_i(1 - N_i) - \alpha N_i P/(1 + \beta N_1 + \beta N_2) \tag{6.5}$$

$$dP/dt = P(((\alpha N_1 + \alpha N_2)/(1 + \beta N_1 + \beta N_2)) - D) \tag{6.6}$$
$i = 1, 2$ notify two potential prey species.

N_i is the population size of prey species i, P is the population size of the predator, α is the consumption rate and β the handling time associated with the predator. D gives the energy/nutrient requirement of the predator. These equations produce a range of cycles in the population densities of N_i (Fig. 6.14). Depending on parameter choice one can get limit cycle oscillations differing in period and amplitude.

Figure 6.14 illustrates that even though both species have the same characteristics, the addition of a second species results in increased oscillations in the abundance of the prey, which leads to higher instability. A single species system is more stable and is unlikely to become extinct even though the mean average density is 16% lower compared with the two-species system. For ants this may mean that potential prey/partners will pass through large changes in density, which undoubtedly will affect their interaction with these partners.

Figure 6.15 is an output of the model showing the relative change in the mean density of one partner of ants when a second species, which has the same

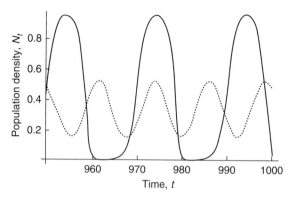

Fig. 6.14. Examples of population densities of prey species 1 based on equations (6.5) and (6.6). Single species system (dotted line) and a two species system (solid line), when the two species have the same characteristics. Parameters are: $\alpha = 5$, $\beta = 3$, $D = 0.8$. α, consumption rate; β, handling time; D, energy/nutrient requirement of the predator. (After Abrams *et al.* 1998.)

attributes, is removed, assuming different energy/nutrient requirements D of a predator. For example, each of the three panels assumes a different value for the consumption rate of ants (α), which might change over the season according to the energy/nutrient requirement of the brood (D).

The simulations suggest that when there are cycles in abundance (Fig. 6.15) apparent competition is reduced for all parameter values. For some parameter combinations the indirect interactions between prey/partners of ants are mutualistic (negative values indicate mutualistic interactions; that is, a decrease in the mean density following removal of the second species), while for others the interaction is antagonistic (positive values = increasing change in population density). Small handling times (β) do not result in mutualistic indirect interactions and relative population change will steadily increase with increasing energy requirement (D). Similarly, if the conversion efficiency (α) of the ants is small, apparent mutualism is unlikely and if it occurs it will only be over a narrow range of D values. Generally, the larger the energy requirements of the ants, the less likely there will be positive indirect effects between partners of ants. Also, the strongest apparent mutualism occurs when there are major differences between the properties of the population cycles in single and two-prey systems; large cycles reduce apparent competition. It is easy to extend this model to situations where the prey/partner growth rates are not logistic but follow some other function or change the predators' per capita growth rates (Abrams and Matsuda 1996). However, as not enough is known about the nonlinearity of functional and numerical responses and prey/partner density dependent growth rates to estimate the probability of apparent competition

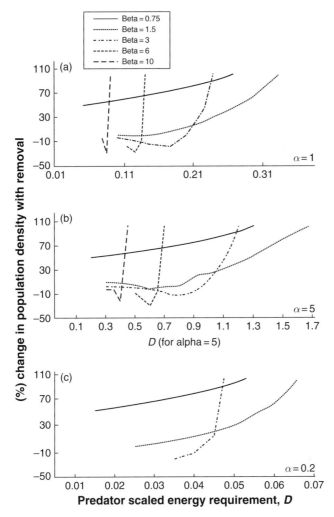

Fig. 6.15. Relative change in population density after one species of prey is removed from the model, based on Eqs. (6.5) and (6.6). Positive values indicate apparent competition, while negative values indicate mutualistic relationships between ants and their partners. Simulations are given for different consumption rates of ants: (a) $\alpha = 1$; (b) $\alpha = 5$; (c) $\alpha = 0.2$ depending on increasing energy/nutrient requirements, D. Lines represent different handling times, β, required by the ant to process prey or mutualist. Note the different scales on the x-axes. (After Abrams *et al.* 1998.)

versus mutualism, we do not pursue this further but look for experimental support for the existence of apparent competition/apparent mutualisms in systems with large population cycles.

A good example of organisms that show marked population cycles are aphids. They undergo large population fluctuations during a season, with an early increase in numbers and subsequent decline due to dispersal or natural enemies. Apparent competition between two aphid species may arise if one species increases in abundance and, as a consequence, negatively affect the abundance of another species because of an increased response of shared natural enemies. In attempting to test this hypothesis Müller and Godfray (1997) manipulated colonies of the nettle aphid (*Microlophium carnosum*) and compared their performance on potted nettle plants in fertilized grass plots infested with large numbers of *Rhopalosiphum padi* and unfertilized plots with few *R. padi*. In those plots where *R. padi* was abundant the nettle aphid suffered an earlier population decline and produced fewer winged morphs than control colonies in unfertilized plots. The reason for the earlier population decline of *M. carnosum* is that natural enemies (larvae of *Coccinella septempunctata*) appeared about one week earlier on nettles in the fertilized plots and in higher numbers than in the unfertilized plots. Thus, *R. padi* had a negative indirect effect on *M. carnosum* via its shared natural enemy.

Extending this experiment to a larger community of aphids comprising 28 species, some of which are ant attended, Müller and Godfray (1999) provide evidence that the composition of aphid communities is influenced by diffuse apparent competition between different aphid species subject to natural enemies and competition for the services of ants. A similar conclusion was reached by Bishop and Bristow (2001) in a study conducted in jack pine forests in the Rocky Mountain region. They showed that the outcomes of apparent competition between aphids and soft scale insects are mainly associated with the dominant ant species (*Formica exsectoides*). The presence of large populations of aggressive, honeydew-seeking ant species can shift the homopteran community from one composed primarily of non-myrmecophilous to one composed of myrmecophilous species. Other good examples of how interactions are affected by a third party and may thus turn from positive to negative or vice versa come from ant–plant–homopteran interactions (Gaume *et al.* 1998, 2000), aphid–plant interactions via host manipulation (Petersen and Sandström 2001), or the interaction between *Maculinea* butterflies and their associated ant hosts via habitat characteristics (Hochberg *et al.* 1994).

The latter example is particularly instructive in the way indirect interactions can act on trophic food webs and may range from apparent competition to apparent mutualism depending on environmental heterogeneity. The univoltine butterfly *Maculinea ribeli* has a complex, parasitic life cycle (Fiedler 2001), which has an impact on several other species. A general diagram of the interacting species is given in Fig. 6.16. Typical sites, which support this community, are subalpine meadows with boundaries defined by the distribution of *Gentiana cruciata*, which is the early larval food plant of the butterfly.

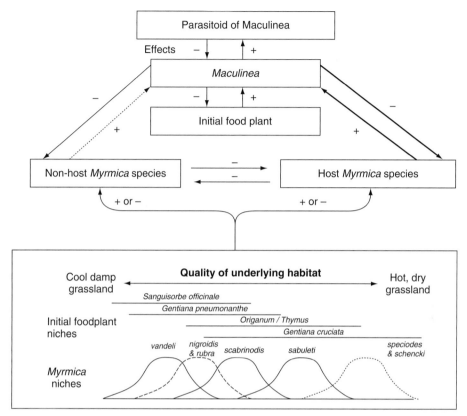

Fig. 6.16. Diagram of the interactions between the species in an ant–plant–lycaenid food web. + indicates beneficial, – indicates negative effects on a particular recipient species identified by an arrow. Boldness of the arrows indicates the relative strength of this relationship. Bottom panel gives the niche breadth of *Maculinea* food plants and *Myrmica* ants over a range of different environmental conditions, which define the quality of the habitat. (After Thomas *et al.* 1997.)

After oviposition in July, eggs hatch quickly and the emerging larvae feed for 2–3 weeks in the flower buds. On reaching the fourth instar, caterpillars fall to the ground and await discovery by *Myrmica* workers, which carry them to their nest apparently mistaking them for ant larvae. The caterpillars cannot disperse more than a few centimetres from the gentian and will eventually die if they are not found within the short foraging range of a *Myrmica* colony. Inside the nest they feed for about 10 months on the brood of the ants before pupating. Although *M. rubra*, *M. scabrinodis* and *M. sabuleti* will carry the larvae to their nests they only survive in nests of *Myrmica schencki*. *Gentiana cruciata* grows in a wide range of conditions from moist meadows, where *M. rubra* and *M. scabrinodis* are dominant, to sparse dry turf, where *M. schencki* is the most abundant species. Ant workers usually forage in the vicinity of their nests (up to 1.5 m). Thus, if a colony occurs within 1.5 m of a flowering *Gentiana* spp., the butterfly can potentially considerably reduce the growth rate of the ants, especially *M. schencki*. Consequently, the density of *G. cruciata* is an important feature determining the quality of a site for this ant.

Different *Myrmica* species with overlapping niches compete for nest sites. Many field experiments have shown that *Myrmica* populations respond rapidly to successional changes in the structure of the vegetation, with thermophilous species replaced by those that prefer cooler habitats within 2–3 years (Vepsalainen *et al.* 2000). This could be because less grazing results in lower temperatures at the soil surface. The *Gentiana–Myrmica–Maculinea* community is the dominant community in these habitats and because of this one may expect clearer patterns of interactions in communities that are less insulated from the influence of other species.

A series of studies in the French Alps showed, for example, that workers of *M. schencki* are consistently less attracted to baits under gentians compared with baits 3–4 m away from a flowering plant (Thomas *et al.* 1997). Excavation of *M. schencki* colonies at several sites and over more than 12 years showed that the average size of a colony in the vicinity of gentians is about half that of colonies further away. In contrast other *Myrmica* species have equal-sized colonies close to and far away from gentians. In addition to this, turnover rates of *M. schencki* colonies near gentians were significantly higher over a five-year period than those for other *Myrmica* species (Thomas *et al.* 1997). Other species of ants, which compete with *Myrmica* to some extent but are not exploited by *M. rebeli*, even tended to increase in the vicinity of gentians. This is assumed to be the consequence of vacated nest sites being colonized by these species. These interactions may be an example of 'weak' apparent mutualism between gentians and non-*Myrmica* or *Myrmica* species other than *M. schencki* because ants benefited from vacated sites and plants benefited from the reduced survival of the lycaenid larvae in ant nests to which they are not closely adapted. The situation with the less suitable *Myrmica* hosts is, however, more complex if the range of habitats where *G. cruciata* can occur is taken into account. For example, in wetter habitats where *M. schencki* does not occur the lycaenid larvae can harm other *Myrmica* species, which would suggest the existence of apparent competition between ant species other than *M. schencki* and gentian plants. A model prediction of positive and negative indirect interactions in this system is given in Fig. 6.17. Here the extent to which direct damage by the lycaenid is compensated for by its greater or smaller impact on the competing ant, *M. schecki*, is important.

For example, in the coolest habitat only species of *Myrmica* that do not host the butterfly can persist as the net effect of the butterfly on the other *Myrmica*

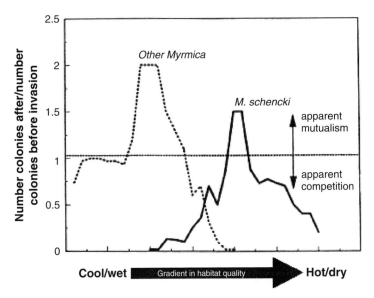

Fig. 6.17. Variation in the outcome of indirect interactions between *Gentiana cruciata*, *Myrmica schencki* and other *Myrmica* species suggested for a subalpine meadow community. Apparent mutualism and apparent competition are possible interactions between the plant and different ant species depending on habitat quality, which ranges from cool wet to hot dry sites, where this species of gentian grows. *Y*-axis represents the mean number of ant colonies in different habitats after and before the lycaenid invaded the site. A ratio larger than 1 means that more butterflies are invading a site and may change the indirect interactions from positive to negative and vice versa for particular *Myrmica* species. (After Thomas *et al.* 1997.)

species is harmful. Thus, apparent competition is predicted between ant colonies that are within 1.5 m of gentian plants, because they suffer at least some brood mortality. In somewhat drier habitats *Myrmica* species that do not host the butterfly compete with *M. schencki* colonies for nest sites and the latter species is usually eliminated from such habitats by *M. rebeli* leaving vacant nest sites, which may be eventually colonized by other *Myrmica* species. So, even though non-hosting *Myrmica* colonies are also damaged by the butterfly they are indirectly benefited by the greater harm done to their competitor, *M. schencki*. The net effect is an increase in the number of other *Myrmica* colonies in this environment. A smaller population of *M. schencki* also benefits the plants because they suffer less defoliation by the butterfly larvae. The outcome of this interaction under these environmental conditions may result in an apparent mutualism between the ants and the gentian. Finally, hot, dry habitats are ideal for *M. schencki*, but less so for the other *Myrmica* species, which have a low competitive ability, and so even minor negative effects of the butterfly larvae make them even less effective competitors and unable to colonize the vacant *M. schencki* nest sites. Under these conditions the net effect indicates apparent competition between the gentian and other *Myrmica* species. This complex web of direct and indirect species interactions indicates that the relative impact of one species on another varies depending on where they occur in an environmental gradient or mosaic of habitat patches. Depending on the conditions very different community patterns might be observed.

There is, however, a problem with this simple dichotomy of indirect positive and negative interactions because it tends to classify net positive effects as mutualism, which might not adequately characterize the underlying dynamics and selection forces. Just like direct mutualistic interactions apparent mutualism can be seen as a form of exploitation, which might be difficult to distinguish from cheating (Fig. 6.18).

For example, attendance by ants often results in protection against natural enemies. Thus, if individuals of species 1 provide honeydew/nectar in order to attract ants they also have to pay the price for doing so (Section 5.2). Individuals of species 2 can exploit the protection services by feeding in the vicinity of attended colonies where there are few natural enemies and not pay the price for attracting ants. Thus, there may be cheaters. However, depending on the perspective different labels can be attached to these direct and indirect interactions. For example, if the feeding activity of species 2 positively affects individuals of species 1, either by an Allee effect, induced changes in plant quality (Petersen and Sandström 2001), or reduced susceptibility to natural enemies (Bergeson and Messina 1997, 1998) due to dilution effects, these interactions could be classified as either: (a) apparent mutualism between

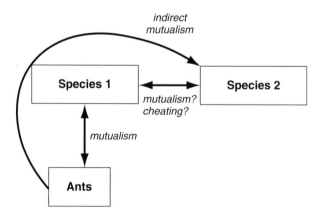

Fig. 6.18. Diagram of the interaction between ants and potential mutualists or cheaters. Interactions might be difficult to classify because net positive and negative effects might vary substantially over time and in different environments.

ants and species 2 (because the presence of species 2 also positively affects species 1, the mutualist of the ants), (b) direct mutualism between ants and species 1, (c) apparent mutualism between species 1 and species 2, or (d) species 2 cheating species 1, by taking advantage of the protection function of ants but not paying for it in honeydew/nectar. Many sap-feeding communities, where ant-attended and unattended species co-occur on the same tree/shoot, might be ascribed to one or more of these categories. Most likely the relative strength of these direct and indirect interactions in a trophically structured community varies considerably in time and space and with the density of the partners, and the net costs and benefits will determine whether the outcome ultimately will be exploitative, deceitful or mutualistic.

IN CONCLUSION, any attempt to understand dynamic community processes and condition-dependent community patterns requires a detailed knowledge of the life cycles and pair-wise effects of the associated organisms, for example the trade-offs in life-history traits, costs and benefits and ultimately fitness changes of each associated member when interacting. This is fundamental to understanding the trade-offs between positive and negative direct and indirect selection forces and the structure of ecological communities to which not only predator–prey (Murdoch *et al.* 2003) or host–parasitoid (Bonsall and Hassell 1997, Bonsall *et al.* 2004) relationships contribute, but also an infinite variety of mutualisms. It is anticipated that this will only be achieved in relatively few communities, but ants and their insect partners might be one such assemblage of species in which it occurs.

6.5 Metamutualism

As said before, population regulation is a central concept in ecology, including evolutionary ecology. By definition, a population is regulated if it shows three closely related features: (1) persistence; long-term survival over many generations, (2) boundedness; fluctuation within certain limits, and (3) the tendency to return to previous levels after a disturbance (Murdoch 1994, Turchin 2001, Berryman *et al.* 2002). As boldly suggested by Turchin (1999) in his principles of population regulation, ecologists believe that populations persist because of the following.

(1) They experience some form of density dependent feedback. Density dependence does not need to operate continuously to regulate a population (Wiens 1977), but it is essential that it operates at some time and place for long-term persistence. This does not mean that density dependence is easy to identify. Even in long time series it is still difficult to distinguish a stochastic but regulated population trajectory

from unregulated random walk (Shenk *et al.* 1998) and it is still difficult to detect population regulation in nature (Murdoch 1994).

(2) Density dependence is necessary but not sufficient for population regulation. It must also be sufficiently strong to counteract any potential disruptive events from density independent factors and any time lags must not be so great as to cause destabilizing population cycles that drive the population to extinction. Density dependence does not always ensure persistence. If density dependent processes become weaker relative to disruptive events or stochastic variation then a population can be driven to extinction.

(3) Competition and predation/parasitism are potential mechanisms for density dependent regulation. Competition for a limited amount of resources is always density dependent by definition as a growing number of individuals either directly (e.g. via interference) or indirectly (e.g. exploitation) fight for the diminishing resources. Predation (broadly defined) does not always need to cause density dependent prey mortality, because this requires that natural enemies exhibit a combination of a numerical response to prey population size and/or functional response in the per capita consumption rate in order to have a negative impact on the growing prey numbers. This is often not the case.

As indicated above during the development of theories of population regulation theorists have almost completely ignored mutualistic associations and shown little interest in how these interactions or the associated dynamics influence population persistence, boundedness and return tendency. This equally applies to new developments like the metacommunity concept, which it is thought will provide a framework for multiscale community ecology (Mouquet and Loreau 2003, Leibold *et al.* 2004). Here the new concepts are described and then a metamutualism perspective is added before some case studies including mutualistic relationships between ants and their partners are highlighted.

Given that much community theory focuses on a single scale, assuming that local communities are closed and isolated, considerable progress was made by addressing interactions that occur at other scales (McArthur and Wilson 1967, Levin 1992, Hanski 1998, Maurer 1999, Hubbell 2001). In the course of further developing the idea of interacting entities the metapopulation perspective was extended to the metacommunity perspective (Wilson 1992, Holt 2002, Mouquet and Loreau 2002, 2003) and even metaecosystems (Loreau *et al.* 2003). Metacommunities are defined as a set of local communities that are linked by dispersal of many potentially interacting species (Wilson 1992). Leaning strongly towards the metapopulation concept this idea distinguishes between two community traits defined by space. At the local level classic species interactions, including Lotka–Volterra type models, as well as their more elaborate developments (Section 3.3.1) are included. At the regional level, dispersal of individuals among local communities occurs. Depending on the extent of dispersal either processes at the level of local communities or interconnected metacommunities will dominate. Because of the close similarity with metapopulation theory, metacommunity thinking is based on the same terminology but with additional terms that are defined in Table 6.3 (Leibold *et al.* 2004).

The concept is hierarchically structured consisting of three nested scales. The smallest site, a microsite, can hold a single individual. Microsites are nested in a patch, which includes local communities and many of these patches are connected to a meta-community occupying a region. Microsites, patches or regions are not fixed spatial entities but can be adjusted to the specific properties of the landscape. Four broad paradigms of metacommunity theory are suggested (Table 6.3, Fig. 6.19) (Leibold *et al.* 2004). Some touch upon concepts already described above. They were termed *patch-dynamic*, *species-sorting*, *mass-effect* and *neutral* paradigms. Here these concepts will be addressed one at a time to determine whether they might be useful for describing communities that contain mutualists, in particular ants and partners of ants.

The *patch-dynamic paradigm* includes the competition–colonization trade-off and involves two species competing for empty space in a fragmented setting. Species A is the superior competitor and species B is the superior colonist (Fig. 6.19a). There is one empty patch in this metacommunity, which could be colonized by either species (most likely by B in this case).

This simple scenario also adequately characterizes mutualistic associations because dispersal of partners of ants is relatively restricted compared with dispersal of unattended partners, while their mortality rates are often lower (see examples above). Unattended species have a more fugitive strategy and are likely to be able to reach more microsites compared with unattended patches, and, as a consequence, their local population sizes in any single patch might be low and make density dependent processes less important – but many patches will be occupied. The assumption of the patch-dynamic model is that the local sites do not differ in abiotic and biotic conditions, except for the species composition that exists at any given moment in time. A second important assumption for the competitive metacommunity is that there is sufficient variation in competitive ability and that the trade-off with dispersal is sufficiently negative to permit regional persistence. Again, these assumptions might hold, for example, for ant–homopteran and ant–lycaenid relationships because there are at least temporal hierarchies in the preferences of ants for certain species of aphids (Völkl *et al.* 1999, Fischer *et al.* 2001), membracids (Messina 1981, Bristow 1984) and lycaenids (Fiedler 1997b). Non-dispersing

Table 6.3. *Key terms used to define the metacommunity concept*

Term	Definition
Scales of organization	
Population	All individuals of a single species within a habitat patch
Metapopulation	A set of local populations of a single species that are linked by dispersal
Community	The individuals of all species that potentially interact within a single patch, local area or habitat
Metacommunity	A set of local communities that are linked by dispersal of many interacting species
Description of space	
Patch	A discrete area of habitat, separated by uninhabitable areas from other patches
Microsite	A site that is capable of holding a single individual. Microsites are nested within patches
Locality	Often used in a way analogous to the term patch. An area of habitat capable of holding a local community
Region	A larger area of a habitat capable of supporting a metacommunity
Dynamics	
Colonization	A mechanism of spatial dynamics in which populations become established at sites from which they were previously absent
Dispersal	A mechanism of spatial dynamics involving movement from one site to another
Extinction	A mechanism whereby established local populations cease to exist
Community structure	
Open community	A population/community that experiences immigration and/or emigration
Closed community	An isolated population/ community receiving no immigrants and producing no emigrants
Metacommunity paradigms	
Patch-dynamic perspective	A perspective that assumes that patches are identical. Patches may be occupied or unoccupied. Spatial dynamics are dominated by local extinction and colonization
Species sorting	A perspective that emphasizes the resource gradients with patch quality and dispersal jointly affecting local communities. Dispersal is important for tracking changes in local environmental conditions
Mass-effect perspective	A perspective that focuses on the effects of immigration and emigration on local population dynamics
Neutral perspective	A perspective in which all species are similar in their fitness. Population interactions among species are random and alter the relative frequencies of species.

Note: Competition is not a primary mechanism defining the population dynamics of metacommunities.

Modified after Leibold *et al.* (2004).

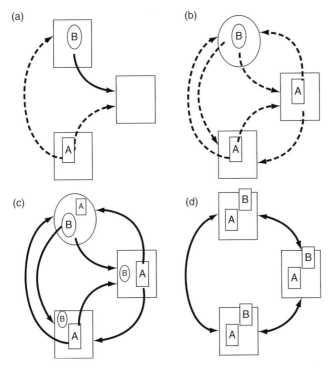

Fig. 6.19. Schematic representation of the ideas incorporated in meta-community theory. In the simplest form a community consists of just two species. There are two competing populations representing species A and B. Arrows connect a site occupied by a species to other sites that can be colonized. Solid lines indicate higher dispersal rates than dashed lines and movement can be either unidirectional (single headed arrows) or bidirectional (two headed arrows, in panel d only). The degree to which one species is competitively superior at a site is given by the size and locality of the smaller box or oval (denoting its habitat niche). (a) Patch-dynamic paradigm, (b) species sorting, (c) mass effect, (d) neutral. (After Leibold *et al.* 2004.)

aphids, for example, will be removed from patches that they share with a species higher up in the preference hierarchy. The inferior species can coexist if it manages to settle on a neighbouring plant or parts of a plant.

The *species-sorting paradigm* incorporates the idea of environmental change acting on communities, with species adapting to specific local conditions over time. The outcome of an interaction is largely shaped by the specific environmental conditions in local patches, assuming that mosaics of environmental conditions exist (Fig. 6.19b, with dashed lines representing a limited degree of dispersal). This assumes that dispersal is not sufficient to alter their distribution. Hence, the follow-up assumption is that mostly specialists inhabit any given patch and that there is no or only a negligible trade-off with dispersal.

Consequently, there is a good correspondence between local species composition and abiotic conditions. In spite of the fact that many phytophagous insects are highly host specific and mutualism between ants and their partners requires morphological, physiological and behavioural adaptations (see Chapter 4) this paradigm provides a less suitable explanation of the mutualisms between ants and their partners. Most partners of ants go through a highly mobile stage or show alary polymorphism, which means that at least at some time during their life cycle dispersal leads to a strong exchange of species between local communities. Local sorting, however, might then quickly eliminate those species that are less adapted to the local conditions to which the winged morph was originally adapted. However, such a high degree of specialization to local conditions is unlikely if there is much gene flow.

The *mass-effects paradigm* assumes that all species in a meta-community have high dispersal abilities (Fig. 6.19c). Here different patches are assumed to provide different conditions at a given time and they are sufficiently connected to each other that dispersal can result in resource–sink relationships between populations in different patches. Immigration and emigration can either increase or decrease local population sizes beyond that expected on the basis of dynamics prevailing in closed populations. At very high dispersal rates coexistence is likely to be reduced because there is little variance in fitness, potentially leading to a homogenization of the communities. In contrast, the patch-dynamic paradigm assumes that there is sufficient variance in competitive ability and that species adapt to different environmental conditions. Mass effects cause species to be present in both source and sink habitats, however, with different population sizes. In a meta-community setting it is unlikely that all species have the same dispersal abilities (indicated by the identical width of the arrows in Fig. 6.19c) but given that, for example, aphids, membracids, coccids and lycaenids are capable of long-distance dispersal it is conceivable that such a pattern might arise, at least temporarily, in communities consisting of ants, myrmecophiles and non-myrmecophiles.

In the *neutral model*, the assumption is that there is no variation in life-history traits and consequently no co-variation, for example with environmental conditions (Bell 2001, Hubbell 2001). Contrary to the above paradigms all species are identical in their niche relationships with local habitat characteristics and ability to disperse. Many experimental ecologists find this difficult to accept, but it is a perfectly good assumption; in particular this neutral view provides a null hypothesis against which any deviations in real systems can be evaluated. Like the patch-dynamic the neutral model does not assume that the conditions at local sites differ, except in species composition. Neutral models have not been applied to communities including mutualists, not to speak of mutualistic associations between ants and their insect partners.

The different facets of the metacommunity approach are useful as they try to categorize complex relationships in a set of predictions and seek principles that can explain ecological patterns at larger scales. While it is strongly based on classic studies, such as island biogeography or metapopulation theory, it attempts to integrate community and population ecology to understand the dynamics and structure of metacommunities. However, in its present state of development it tends to neglect achievements and insights, which have been made at the local scale. The incorporation of top-down or bottom-up effects, competition, density dependence, temporal variation in local and regional processes and the importance of mutualistic associations for the stability and diversity of metacommunities is either completely missing or credited only a minor role in the overall picture. In contrast, it is argued here that this information needs to be incorporated in any paradigm of metacommunity theory and in accordance with our topic focus on mutualistic metacommunities building on the achievements that were addressed in previous chapters. Four broad sets of mutualistic metacommunities are envisaged, which consist of ants, partners of ants and potential cheaters in a spatial context (Fig. 6.20a-d). That is, the metamutualism concept incorporates top-down, bottom-up and spatial aspects. The competition–colonization paradigm is the central underlying theme.

Figure 6.20a shows the situation in which an obligate myrmecophile (B) receives and provides benefits for ants. The facultative myrmecophile (C) also receives and provides benefits but less so compared with the obligate myrmecophile. With increasing population size of B and C the benefits to C will most likely decline as the ants obtain more of their benefits (e.g. sugary excreta) from their obligate partners. As a consequence, the facultative myrmecophile might experience two negative effects: (1) displacement by the competitively superior species B (vertical dotted line) and (2) as a consequence of having eventually to leave, suffering high mortality during dispersal. In addition, during the shift in attendance from both species to only species B, species C is likely to experience more severe top-down effects, again forcing individuals to disperse to other patches.

The situation depicted in Fig. 6.20b is similar to that in Fig. 6.20a and consists of a metacommunity of facultative and obligatorily myrmecophile mutualists with ants, but in this case the ants are limited by a factor other than their mutualist, which could be a top-down or bottom-up process (e.g. competitively superior ant species, unsuitable habitats), indicated by the bold circle. As a consequence, the benefit conferred on the obligate myrmecophile

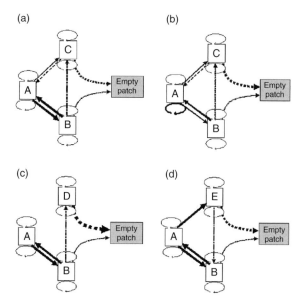

Fig. 6.20. Schematic illustration of the metamutualism concept. A, ants; B, obligate myrmecophile; C, facultative myrmecophile; D, unattended species; E, cheater. Bold arrows indicate strong effects (benefits), thin arrows indicate weak effects (which might even turn into costs). Dotted lines indicate dispersal and the circle below and above a species that inhabits a patch represents bottom-up and top-down effects. Each mutualist is subjected to these two types of vertical effects. For example in panel (b) ants are limited by resources other than their trophic partners. Note that the within-patch processes are connected to between-patch processes because oval arrows (within-patch processes) are connected to straight arrows (inter-patch processes).

B declines, while the benefit of species B to A and the relationship between A and C remain unchanged. This also means that B is now less likely to displace C and might also have a lower probability of colonizing other patches. Obligate myrmecophiles are likely to directly experience more of the constraints that act on their partners than facultative myrmecophiles because they are more intricately associated with the ants. In this mutualistic meta-community limitations imposed on the ant partner might facilitate the coexistence of different myrmecophile strategies.

In a patch where unattended (species D) and obligate myrmecophiles (B) co-occur, D does not receive any benefits from ants or suffer any direct costs. Indirect costs, however, might be considerable because the mutualists A and B might quickly usurp the available local resources, resulting in the unattended species migrating to other patches. Unattended species are most likely to be the most mobile (thickest dotted arrow). Figure 6.20c most closely mirrors the classic competition–colonization trade-off in which coexistence of the species is only possible if D is the better disperser and B the better competitor.

Mutualistic interactions are always at the risk of being exploited by cheaters (E) who receive benefits from ants but do not reciprocate (Fig. 6.20d). For example, for ants some aphids, coccids or lycaenids are unpalatable and provide no honeydew or nectar. Their close proximity to attended species might confer protection against natural enemies but they do not pay for the protection. There are a number of examples that might support the conclusion that exploiting a system like the one described here might indeed play a role in these metacommunities. Because the cheater benefits from ant attendance and pays no costs, growth rates might even be larger than those of the obligate myrmecophile, which eventually could lead to the competitive displacement of B, especially as dispersal of the cheater is likely to be less impeded compared with the obligate myrmecophile, B. The extent to which a cheater might displace a myrmecophile in a mutualistic metacommunity or destabilize the whole metacommunity is unclear, but the implication is that obligate myrmecophiles are unlikely always to do better than other species that have slight or no association with ants. The notion that exploitation is a serious threat to the persistence of mutualists and causes evolutionary instability by eroding mutualistic interactions permeates the literature (Axelrod and Hamilton 1981, Doebeli and Knowlton 1998, Herre *et al.* 1999). However, more recent theoretical analyses for plant/pollination systems clearly indicate that stable coexistence of mutualists and exploiters is possible over a broad area of parameter space, such as birth rates of exploiters and competitive abilities (Ferriere *et al.* 2002, Morris *et al.* 2003).

The outcome of the interactions between ants, unattended species, myrmecophiles and cheaters is highly dependent on the temporal dynamics of the within-patch processes. In any of the above figures the impact of density dependent processes, plant quality effects or temporal variation in top-down and bottom-up forces plays an essential role, which indicates that both negative and positive feedbacks are important at the local scale. Asymmetries in the relative strength of these processes, resulting from density dependent processes or seasonal changes in metabolic abilities and life-history traits, and in the spatial variability in resource availability are prevalent in nature, and provide mechanisms for facilitating coexistence along the antagonistic–mutualistic continuum in metacommunities. An example of such asymmetries is given below.

The reproductive success of group-living insects depends on their ability to locate, exploit and defend food resources. In ants foraging efficiency is achieved by a small number of scouts directing the workers to the resources via marking trails (Hölldobler and Wilson 1990). Central place foragers, like many ant species, do not evenly operate in all sectors of their territory but instead exhibit asymmetries in their use of foraging space, which basically reflects the heterogeneity of their environment, such as the spatio-temporal availability of food resources (Brown and Gordon 2000). Mutualistic relationships require, by definition, close contact of the partners and because migration in insects is costly and risky (Rankin and Burchsted 1992, Ward *et al* 1998) social insects allocate individuals to find the resources they need. In social insects, like ants, the workers can be asymmetrically distributed, even in a uniform environment. This self-organizational process (Bonabeau *et al.* 1997, 1999, Portha *et al.* 2002, 2004) may account for complex spatio-temporal features that emerge at the colony level. For example, *Lasius niger* colonies containing brood mobilize more workers for collecting sucrose than for collecting protein. This suggests that the location of carbohydrates elicits a higher trail-laying behaviour than protein food does. Sugar is therefore a key element in the mass-recruitment behaviour of *L. niger* as it indicates a highly rewarding resource such as an aphid colony.

Foraging patterns also differ according to food type: individuals focus their activity on only one droplet of sucrose, whereas the colony foragers are rather homogeneously distributed between proteinaceous sources (Portha *et al.* 2002). For mutualists of *L. niger* this means that a colony of ants avoid dispersing workers between multiple sources and receiving the services of a colony might only be possible in the close proximity of an ant colony. Searching for the more scattered proteinaceous sources, in contrast, never leads to mass recruitment of workers, a behaviour that might maximize the probability of discovering small prey items. Ant colonies also adjust their harvesting strategy to the internal demand for nutrients within the nest and the number of workers allocated to each food type reflects the nutritional needs of a colony (Sudd and Sudd 1985, Fowler 1993). Thus, the asymmetry in food exploitation is influenced by internal requirements and the effect this has on the dispersal strategies of potential partners of ants because their environment is a mixture of potentially high-quality patches where ants are present and low-quality ones where ants are absent. These observations suggest that an adequate theory of metamutualism must explain not only extreme reciprocal adaptations but also the existence of substantial local maladaptation. Below are presented a number of examples that highlight the different processes structuring a metamutualistic community, indicated in Fig. 6.20.

IN SUMMARY, population regulation and community effects are not adequately understood if positive interactions between organisms are ignored. What is required is a detailed understanding of how positive and negative effects at a local scale (competition) interact with dispersal (colonization) at a regional scale. To achieve this goal a multidisciplinary approach is necessary.

6.5.1 Experimental evidence for metamutualism

Recent work in this direction has been stimulated by Thompson's geographic mosaic theory of coevolution (Thompson 1994). This theory suggests that coevolution will commonly lead to spatially variable patterns of local mala-daptation due to selection mosaics, coevolutionary hot spots and trait mixing (Thompson 1997, 1999, Thompson and Cunningham 2002). These processes may be particularly important within conditional mutualisms, where selection varies between mutualism and antagonism in response to temporally and spatially changing environmental conditions (Bronstein 1988, Billick and Tonkel 2003, Nuismer *et al.* 2003, Mooney and Tillberg 2005). For example, the effect of the ant *Formica obscuripes* on the membracid *Publilia modesta* varied considerably between years (Billick and Tonkel 2003). When surveying ant–membracid associations at five sites near Colorado, USA, they found considerable spatio-temporal variation in ant benefits for nymphs of *P. modesta* (Fig. 6.21).

Both year and the presence of ants significantly affected the number of individuals that matured (Fig. 6.21a). Site also significantly affected the survival of nymphs to the adult stage (Fig. 6.21b). Ant tending significantly increased adult membracid numbers and the effect of ants did not vary among sites. The benefit of ant attendance is clearly protection from natural enemies, which in this case appears to be spiders and adult coccinellids. The mechanisms underlying this temporal variation in the positive interactions between ants and membracids are not entirely clear but variation in salticid spider density or yearly variation in ant foraging patterns are suggested as potential factors. Interestingly, though, the effects of ants on nymphs varied among sites through time; this effect disappeared when only adult numbers were examined. This paradoxical result is explained by the negative correlation between aggregation size in *P. modesta* and proportion of nymphs reaching adulthood. While attended colonies had significantly higher nymphal survival these larger colonies also produced proportionally fewer adults, leading to a negative density-dependent adult production (Billick and Tonkel 2003). In conclusion, in spite of the fact that there were site-specific differences they did not translate into spatial variability in the effects of ants on membracids. That is, the scale of this study was too small to detect site × treatment interactions. Temporal variation was thus the primary factor in the mutualism between *P. modesta* and *F. obscuripes* (Cushman and Whitham 1989).

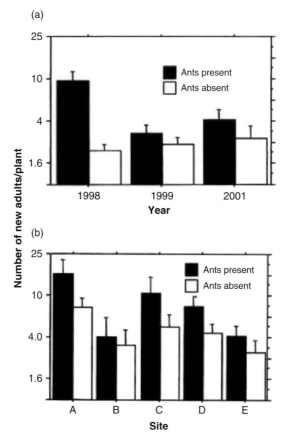

Fig. 6.21. (a) Mean number of adult *P. modesta* produced on the host plant *Chrysothamnus viscidiflorus* over three years at a single site (Site E) at Gunnison National Forest, Colorado, when attended by ants or unattended. (b) Mean number of individuals reaching the adult stage at five sites (A–E), which were located about 0.5–1.5 km apart along a transect. Colonies are separated according to the presence of ants. (After Billick and Tonckel 2003.)

Evidence for spatial effects on mutualistic relationships between ants and their insect partners is common. The concept of ecological neighbourhoods might be useful in explaining the function of spatial units and may help when selecting the appropriate units for studying mutualistic interactions (Fig. 6.22) (Addicott *et al.* 1987).

Ecological neighbourhoods are defined by three properties: (1) an ecological process that determines the neighbourhood to be considered, (2) an implied time scale for that process, and (3) a region of activity or influence during that period of time. Thus, the choice of ecological process will determine an appropriate time scale over which to measure neighbourhood size. Generally, the ecological

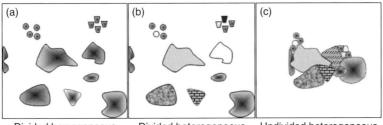

Divided homogeneous   Divided heterogeneous   Undivided heterogeneous

Fig. 6.22. Examples of environmental heterogeneity. Different patch types are marked with different patterns. (a) Divided but homogeneous environment. Populations are either large or small and located at the centre of the patch. Ecological neighbourhood might be experienced as a gradient from the centre of the patch to its periphery. (b) Divided and heterogeneous environment corresponding to a meta-community of one or many different species living together in a patch. (c) Undivided and heterogeneous environment where boundaries between patches are blurred. (Modified from Addicott *et al.* 1987.)

neighbourhood of an organism for a given ecological process (e.g. mutualistic interaction) is the region within which that organism is active or has some effects on other organisms during the appropriate period of time (Addicott *et al.* 1987). Given that an ecological neighbourhood is a transient feature, physical size is just one out of many possible criteria. The mobility of an organism might be equally useful to describe patch size and ecological neighbourhood. For example, for the mainly apterous and ant-attended aphid species, *A. varians*, there is little or no movement from ramet to ramet within a clone of fireweed, whereas the mobile and unattended *M. valeriana* has a much larger ecological neighbourhood (Antolin and Addicott 1991). A field survey showed that *M. valeriana* regularly walks to adjacent shoots of a colony and 95% of the movements were within 75 cm of the central colony, while movements of *A. varians* were usually less than 10 cm and mainly to nearby shoots. This also means that colonies of *A. varians* persisted much longer than colonies of *M. valeriana*. Here the mutualistic interaction with ants is an important process, which defines the ecological neighbourhood temporally. This highlights the necessity to report the scale of ecological experiments and restricts the interpretation of the results across changing scales and associated changes in neighbourhood. The implementation and use of appropriate scaling techniques and the examination of focal systems at multiple spatial and temporal scales seems vital to facilitate comparisons between theoretical postulates and empirical studies, and between multiple empirical studies.

In a similar system consisting of goldenrod (*Solidago altissima*) and the aphids *Uroleucon nigrotuberculatum* and *U. tissotii*, the spatial patterns of the aphid distributions are strongly correlated with their mobility. While *U. nigrotuberculatum* aggregate when colonizing new shoots and the individuals are rather sedentary, producing large colonies, *U. tissoti* is highly mobile and colonizes shoots singly. This different behaviour subsequently resulted in a more random spatial pattern within fields of goldenrod (Cappuccino 1987, 1988). As a consequence of its spatial pattern *U. nigrotuberculatum* is better protected against predators in large colonies and there are fewer predators per unit area in dense patches, but if a fungal pathogen (*Neozygites frensenii*) becomes the dominant mortality agent, aphids in dense patches are more vulnerable than aphids in sparse patches. Thus, when natural enemies act in an inverse density dependent manner (at least temporally) then the dispersion exhibited by *U. nigrotuberculatum* might be more advantageous. However, when faced with a strong density dependent mortality agent, such as a fungal pathogen, a dispersal strategy similar to that of *U. tissoti* is likely to be more advantageous. The effect of natural enemies, such as ladybirds, is strongly dependent on the degree of patchiness because the ecological neighbourhood is interrupted and cannot easily be traversed by natural enemies. By experimentally manipulating the degree of patchiness Kareiva (1987) showed that increasing patchiness leads to more frequent local outbreaks of *U. nigrotuberculatum* and thus less stable dynamics. These results were consistent over four consecutive years. The suggested reason is that increasing patchiness interferes with the non-random searching behaviour of ladybird predators (here *Coccinella septempunctata*) and the specific behavioural response of individuals using this habitat confines them to a smaller ecological neighbourhood. As a consequence of all of this it is reasonable to assume that local within-patch (mosaic) processes define the ecological neighbourhood, which in turn extends to the metacommunity and metamutualism in a fragmented landscape.

Ants show a similar response to the spatial partitioning of their habitat, and are thus also affected by ecological neighbourhoods. A number of studies show spatial and temporal effects in ant communities. For example, Albrecht and Gotelli (2001) studied the activity patterns of entire ant assemblages in ground-foraging grassland ants in Oklahoma. They used monthly 24-hour surveys at tuna-fish bait stations to document the co-occurrence and abundance of ant species. Each month 25 bait stations were placed in a grid pattern at 5-m intervals and the observed frequencies of bait occupancy patterns were compared with null models to examine niche overlap. For tests of temporal niche overlap either hourly or monthly time scales were used and for the spatial niche overlap species occurrence at different bait stations was used. Niche overlap between each pair of species was quantified using the

Czechanovski index (Feinsinger *et al.* 1981), which calculates the overlap between species 1 and 2 according to the fraction of baits visited by either species. The index approaches 0 for species that share no resources and approaches 1 for those that utilize identical resources. Niche overlap patterns for the entire assemblage were calculated by averaging the index over all species pairs in the assemblage. On a seasonal (monthly) time scale niche overlap was found to be either random or aggregated, probably in response to the thermal requirements of the species when foraging in a seasonal environment. In the warmer months of the year there is some evidence for niche partitioning and the strongest evidence for niche partitioning is at the scale of individual baits (Fig. 6.23).

Both expected and observed niche overlap varies greatly between ant species and between different months and hours of the census day. In most months and at most times of the day, observed spatial niche overlap was less than expected on the basis of the null model. This difference was particularly strong in June and July but less so during the cooler months. Temperature appears to be an important determinant of activity patterns in ants (Holway *et al.* 2002b), while nest site availability was probably not limited in this grassland community. Instead, spatial and temporal niche partitioning of foraging activity is suggested to be the primary mechanism for coexistence. No species of ant monopolized a bait and created static co-occurrence patterns that were maintained for several hours or days. Instead there was a considerable turnover in species composition, as different species colonized, occupied and abandoned baits over a 24-hour census period. It is possible that competition-colonization trade-offs contribute to the coexistence of these species, as suggested for another ant community invaded by an exotic ant species (Holway and Suarez 2004). Again, this example provides evidence for dynamic ecological neighbourhoods for ants and partners of ants. There was little evidence for seasonal niche partitioning by ground-foraging ants. Instead, partitioning occurred on short temporal and spatial scales in which species were active and used food resources in a mosaic of patches (Albrecht and Gotelli 2001).

With this background on the dynamics of ecological neighbourhoods, both for ants and partners of ants, and its consequences for species co-occurrence, it is interesting to ask how spatial fragmentation (Fig. 6.22) affects mutualistic interactions. Because mutualisms between ants and their partners are based on trophic relationships, theory suggests that species at higher trophic levels tend to be more vulnerable to fragmentation due to small population sizes (Kruess and Tscharntke 1994, Holt 2002, Tscharntke and Brandl 2004). This theory was tested in an experiment with ground-foraging grassland ants in Switzerland. Braschler *et al.* (2003) studied the effects of fragmentation of

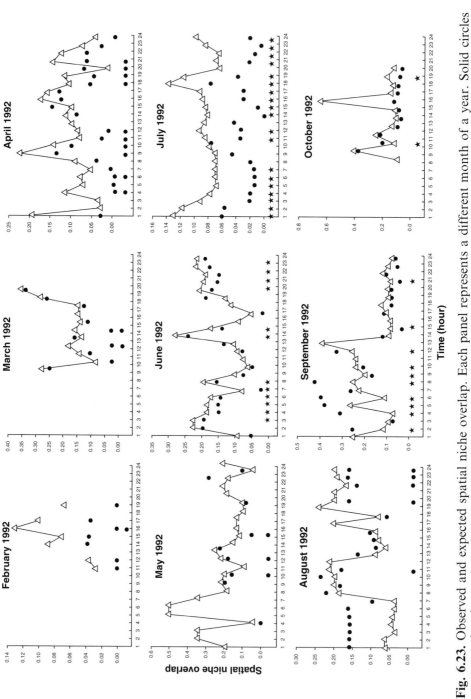

Fig. 6.23. Observed and expected spatial niche overlap. Each panel represents a different month of a year. Solid circles represent the observed spatial overlap among baits calculated at different times of the day; open triangles represent the expected niche overlap; a star indicates an observed niche overlap that was significantly different from the null model. (After Albrecht and Gotelli 2001.)

grassland of different sizes ($0.25\,m^2$, $2.25\,m^2$, $20.25\,m^2$) on aphid abundance, abundance of natural enemies and association with ants. Similar to Kareiva (1987) they found that aphid densities were significantly higher in fragmented plots than in unfragmented control plots. It turned out that this result was a combination of a higher frequency of aphid-infested plants and more ant-attended colonies in fragmented plots. It was suggested that the better services provided by ants in patchy environments, such as the removal of honeydew or protection from natural enemies, contribute to the success of ant-attended species. However, unattended colonies are also more numerous in fragmented habitats, which indicate that fragmentation per se is more important for the establishment of mutualistic associations than the services supplied by ants. The suitability of a habitat for ants in temperate regions is mostly defined by temperature and humidity, with most ant species preferring relatively warm and dry localities (Gösswald 1938, Hölldobler and Wilson 1990). Both factors were affected by experimental grassland fragmentation: temperatures increased at the edges of fragments and thus became drier (Zschokke *et al.* 2000, Braschler and Baur 2003), and this is where ants tended to build their nests. One ant species became particularly abundant (*Lasius paralienus*) and its dominance was negatively correlated with species richness of other ant species in fragments but not in control plots. This might be either a consequence of competition for food but more likely competition for suitable nest sites (more warm sites in fragmented plots). Generally, although ant communities are commonly considered to be mainly structured by competition, a high niche overlap with the dominant *L. paralienus* suggests, similar to the conclusion of Albrecht and Gotelli (2001) and others, that ant species partition resources at fine temporal and spatial scales.

Fragmentation changes the abiotic conditions of the microsites (ecological neighbourhood) and increases carbon resources (increased availability of honeydew) and thus may have a twofold consequence: (1) reduction of available sites for subdominant ant species and (2) at least a temporal reduction in the intensity of interspecific competition for sugar resources amongst ants (Braschler and Baur 2005). In addition, persistence of the dominant ant species was shorter in fragmented than in control plots but longer when situated closer to the edges than in the core area, suggesting an opportunistic behaviour and high local turnover rates in this mutualistic metacommunity comprising different ant and aphid species. Parasitoid pressure was not affected by fragmentation but no information is available on predators.

Further evidence for the effects of spatial variability on aphid–ant relationships comes from forest succession producing local fragments of open terrain and almost impermeable patches of dense stands of Norway spruce. For

Table 6.4. *Characteristic monogynous and polygynous wood ant species in relation to successional dynamics and landscape structure of boreal forests*

Environmental features	Monogynous	Polygynous
Forest successional stage	Early	Late
Disturbance frequency	High	Low
Fragment size	Small	Large
Fragment isolation	High	Low
Location of the nest site	Edge	Interior

Colony structure is defined as monogynous, meaning that colonies have just one reproductive queen per nest; polygynous, multiple reproductive queens.
Source: After Punttila (1996).

example, the polygynous ant *Formica aquilonia* is more abundant in old growth forests and large fragments of forests and the monogynous *F. lugubris* in young forests, small fragments of old forests and the edges of forests (Punttila *et al.* 1991, Punttila 1996). Forest fires and local dieback create open, ant-free patches, which are rapidly colonized by thermophilic species, including the *F. fusca* group. This again suggests the existence of a competition–colonization trade-off between species that are good at colonizing ephemeral habitats and those that are competitively dominant, especially if environmental conditions become less favourable for the active and fugitive boreal ant communities. The ecology of ants is suggested to be considerably influenced by the structure of the landscape of boreal forests (Table 6.4).

Monogynous species tend to occupy only one nest per colony (monodomy), while polygynous species tend to form multinest colonies (polydomy). Workers are often less aggressive and smaller than those of monogynous species. Nests of a multinest colony are often linked and, as a consequence, can reach into less favourable habitats than single nest colonies. As a consequence of the presence of multiple queens and multiple mating, individual workers of polygynous species are less related to one another than individuals of colonies with only a single reproductive queen. As described in Section 4.2.1, these colony features can be expected to have considerable consequences for mutualistic partners of ants and provide fertile ground for the mosaic theory of coevolution, which suggests spatially variable patterns of local adaptation/maladaptation due to selection mosaics, coevolutionary hot spots and trait mixing.

A direct consequence for mutualistic associations is that the partners of ants are also indirectly affected by successional processes and landscape structure.

However, given the dominance of spruce in boreal forests Punttila (1996) suggests that boreal ant communities, i.e. those in spruce-dominated forests, are influenced by succession more than the species of aphid they attend. Nevertheless, there is good evidence that ant communities can affect the composition of aphid communities (see apparent competition/mutualism; Section 6.4).

6.5.2 *Examples of multiple mutualistic interactions*

So far we have been dealing with studies that focus on the temporal and spatial aspects of mutualism by using largely pair-wise interactions and staying within certain trophic levels. However, mutualisms between individuals of just two species or trophic levels are the exception rather than the norm. Now studies that pursue a simultaneous application of bottom-up and top-down perspectives of metamutualism across several trophic levels will be stressed. In so doing species assemblages and the potential interactions in local and regional communities, constraints in trophic relationships and competition–colonization trade-offs are described.

As indicated above (Section 6.1) there is a huge variety of abiotic and biotic factors that can influence the abundance of herbivores. Because phytophagous insects constitute an intermediate trophic level, fluctuations in the availability of food, food quality and natural enemies are likely to affect their abundance. As a consequence, the mutualistic partner of a herbivore should also be affected if the abundance or diversity of the plant consumer changes. However, as simple as this argument might seem, there is little information on the relative effects of bottom-up and top-down forces on mutualistic associations, especially in a metacommunity setting. The problems start when trying to define food quality for partners of ants. Historically, herbivorous insects, such as wood-boring scolytid bark beetles, are thought to benefit from plant water stress due to their attack strategies relying on stress-induced decreases in oleoresin pressure, which facilitates the successful attack of conifers (Koricheva *et al.* 1998). Similar field observations made on sap feeders led to the suggestion that drought stress is a major factor causing outbreaks of herbivorous insects (White 1969, Mattson and Haack 1987). As might be expected, this provoked discussions on the importance of density dependence in the regulation of natural populations versus limitation of the environment (Berryman *et al.* 2002, White 2004). However, plant stress can occur in very different forms and affect herbivore performance in diametrically opposite ways. For example, a meta-analysis of the effects of water stress on major feeding guilds of herbivorous insects (sap feeders and leaf chewers) and subguilds of sap feeders (phloem, mesophyll and xylem feeders), and chewing insects (free-living chewers, borers, leaf miners and gall formers) (Huberty and Denno 2004) revealed that most members of the various feeding guilds showed a negative response to plant water stress, with sap-feeding subguilds being much more adversely affected than chewing subguilds (Fig. 6.24).

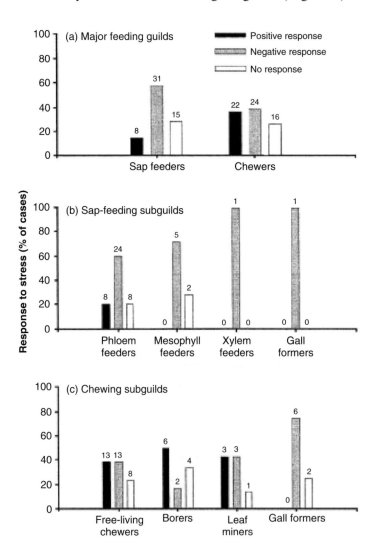

Fig. 6.24. Number of studies showing that plant water stress negatively, positively or does not affect the performance of different feeding guilds: (a) sap feeders and chewing insects, (b) and (c) different subguilds of the major groups. Performance includes survival, fecundity, developmental time, body mass, growth rates, etc. on stressed compared with non-stressed plants. Numbers above the bars indicate the number of observations in each response category within feeding guilds. (After Huberty and Denno 2004.)

Evidence from many laboratory studies suggests that sap feeders are especially adversely affected by continuous water stress of their host plants (Awmack and Leather 2002, Hale *et al.* 2003). This conflicts with several field studies, which reveal that under drought conditions sap feeders, such as psyllids (White 1969) and aphids (Miles *et al.* 1982, Larsson 1989, Larsson and Bjorkman 1993) show increased population growth on water-stressed plants. To resolve this discrepancy Huberty and Denno (2004) proposed a conceptual model to describe the mechanisms that might result in these different outcomes (Fig. 6.25).

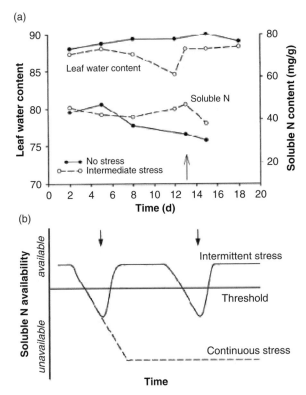

Fig. 6.25. (a) Leaf water and nitrogen content in water-stressed plants. After a period of water stress incipient precipitation (arrow) quickly leads to an increase in leaf water content and turgor pressure, while soluble nitrogen levels may still remain high. (b) Conceptual model of the effects of different intensities of drought stress on nitrogen availability to phloem-feeding insects. Continuous water stress makes nitrogen unavailable to sap feeders, while repeated short-term droughts benefit sap feeders because occasional precipitation (arrows) results in an increase in plant turgor and an elevated nitrogen concentration in the phloem sap. The repeated occurrence of water stress may affect the density of ant partners and impose a high degree of unpredictability on mutualistic relationships. (Data and model after Kennedy *et al.* 1958, Kennedy and Booth 1959, Huberty and Denno 2004.)

When plants experience water stress, protein metabolism and amino acid synthesis are impaired leading to increased levels of free amino acids in the plant tissue (Brodbeck and Strong 1987). In addition, during water stress turgor pressure and water content decrease (Hsiao 1973, Inbar *et al.* 2001). Continuous water stress is characteristic of most experimental studies, whereas recurring wet and dry periods characterize many natural situations. Because phloem-feeding insects require a positive turgor pressure if they are to imbibe sap and extract nitrogen, periods of turgor decline and subsequent increase may be beneficial, while continuous water stress is not. Below a certain threshold, water-stress-induced reduction in turgor may prevent phloem feeders from extracting soluble nitrogen despite its elevated concentrations. With intermittent precipitation turgor pressure will quickly return, while nitrogen concentrations in the phloem are still relatively high. Phloem feeders are therefore likely to benefit temporarily if plants experience intermittent and repeated water stress. The 'pulse-stress hypothesis' proposes that changes in turgor pressure result in an increase in the availability of plant nitrogen and is thought to be the key explanation for the discrepancy between observed outbreaks of sap feeders in the field and their consistently poor performance on constantly water-stressed plants. More generally, plant stress and plant vigour may not necessarily have opposing effects on sap feeders. For example, *Euceraphis betulae* is more abundant on vigorously growing birch branches early in the season than on exposed branches destined to become stressed, but aphids became significantly more abundant on stressed branches later in the season, when symptoms of stress became apparent (Johnson *et al.* 2003b). Therefore, in a plant–herbivore system, where plant quality is highly variable over a small temporal and spatial scale, vigorous and stressed plant parts may be present simultaneously, and hence both plant vigour and plant stress may positively affect the abundance of herbivores. This suggests that regarding these two hypotheses as mutually exclusive might not be appropriate in many plant–herbivore relationships.

Relative to the metamutualism perspective one can ask how repeated bouts of (water) stress might affect the interactions between ants, their insect partners and ultimately community structure? Water stress might not be uniformly experienced by plants in the same region or even in the same habitat. Small changes in elevation or topography could make a considerable difference to how long plants and their associated phytophagous insects might experience drought conditions. Therefore, a reasonable assumption is that adaptive and plastic responses of mutualists to variation in the quality of their partner and environmental conditions are likely to play an important role in determining the dynamics of these mutualistic metacommunities. If repeated periods of stress enhance variability in microsite quality for phytophagous insects, then they should remain mobile so that they can find the sites which temporarily provide the best conditions for maximum growth rates. Thus, under these conditions associations with ants may be less beneficial if they are confined to specific feeding sites close to an ant nest or to specific plant organs. For example, on trees with a mosaic of leaves, shoots and branches in different states of development, water stress or exposure to sunlight or wind, a high degree of mobility should be maintained to access preferred microsites. This is what is observed in many insects (Roff 1994, Dixon 2005). Plant water stress might also affect the production of honeydew, which might be less on wilted plants or plant organs, but it is unknown how this affects the mutualism with ants.

There is a clear shortage of studies that address community-level effects of mutualistic associations from both a bottom-up and top-down perspective. One exception is the study of Wimp and Whitham (2001) who examined the factors that affect the distribution of the obligate myrmecophile *Chaitophorus populicola* on different hybrids of cottonwood *Populus fermontii* × *P. angustifolia* in Utah, USA. Using an observational and experimental approach they studied the relative and interactive effects of top-down (predators) and bottom-up (hybrid types) forces on aphid distribution, fecundity and arthropod community structure on poplar. This aphid is attended by *Formica propinqua*, which builds large colonies and aggressively defends aphid colonies in the vicinity of an ant mound. There is a sharp decline in ant attendance with increasing distance from ant mounds. Consequently, populations of aphids sharply declined with distance, from an average of 1200 individuals per tree close to ant mounds to zero only 8 m from ant colonies. Experiments showed that the decline in ant attendance with increasing distance from the ants was associated with an increase in the exposure of the colonies to natural enemies and decline in the removal of honeydew, which negatively affected aphid fecundity.

In a common garden experiment, using different species and hybrids of poplar, aphid performance varied significantly (Fig. 6.26). Aphid fecundity was lowest on Fremont cottonwood and highest on narrowleaf cottonwood, while intermediate fecundities were recorded on the hybrids. This result was very robust, with identical trends in different years. This was largely corroborated by field surveys at 37 sites along a 500 km long transect of the Weber river, where the geographical distribution of the aphid depended on its differential performance on the different hosts. Because of the dependence of *C. populicola* on particular hosts and ants, the realized aphid habitat is a fraction of the potential habitat. The authors calculated that the interaction between host plant suitability and presence of tending ants limits the distribution of the aphids to only 21% of their potential habitat space.

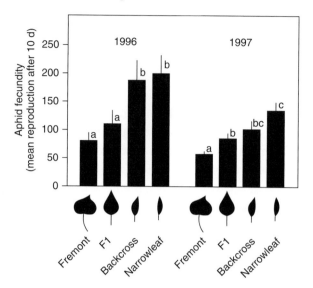

Fig. 6.26. Aphid fecundity on different species (Fremont (*Populus fremontii*), narrowleaf (*P. angustifolia*)) and hybrid types (F1, Backcross) of poplar in a common garden experiment conducted in 1996 and repeated in 1997. (After Wimp and Whitham 2001.)

There was also a clear effect of ant attendance on arthropod community structure. The arthropod community included 90 species in the orders Araneae, Acari, Opiliones, Ephemeroptera, Orthoptera, Dermaptera, Hemiptera, Homoptera, Tysanoptera, Neuroptera, Coleoptera, Diptera and Lepidoptera, and there was a 51% greater species richness and 67% greater abundance on trees, where the aphid ant mutualism was absent relative to those where it was present. The same pattern was found on trees where the mutualism was disrupted by removing the aphid partner. Different guilds, however, were affected differently by aphid removal and the subsequent abandonment of trees by the aggressive ants (Fig. 6.27). Generally on trees from which the aphids were removed the abundance of herbivores increased by 76%. The effect was most clearly seen for herbivores other than aphids and tending ants when the aggressive *Formica propinqua* was removed. As expected, if no aphids feed on the trees, specialized natural enemies of aphids also decrease (44%). As a consequence, the effect of the aphids on the community structure on poplar is indirect via its mutualism with the aggressive *F. propinqua*, which monopolizes its honeydew production.

This study is exceptional, because it demonstrates the strong community-wide impact of plant–herbivore–ant interactions. Mutualistic associations can be the dominant force structuring herbivore communities, because many other

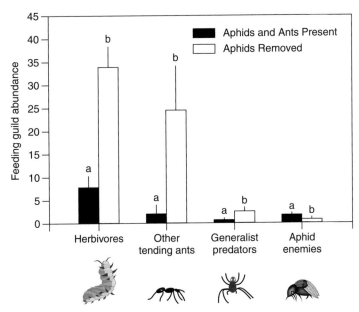

Fig. 6.27. The effect of *C. populicola–F. propinqua* mutualism on the abundance of broadly defined feeding guilds. Different letters indicate significant differences between groups (means + 1 SE). (After Wimp and Whitham 2001.)

herbivores, aphid natural enemies and ants are affected by this key association. Positive top-down effects from ant mutualists, negative top-down effects from predators/parasitoids and a mixture of bottom-up effects due to host plants of different suitability contribute to a complex interaction web, which, when separated into different components, provides a good understanding of the dynamics in mutualistic metacommunities. Similar approaches are adopted by Gonzales *et al.* (2002) and Renault *et al.* (2005), who used a top-down and bottom-up perspective to decipher the relative impact of different trophic levels, interspecific competition via induced plant responses, seasonal changes in environmental conditions and the subsequent effects on population size and mutualist guilds.

6.5.3 Extrafloral nectaries

The honeydew or nectar produced by the partners of ants is not the only carbohydrate resource available to ants. Another interesting bottom-up effect, which could affect mutualistic interactions between ants and their partners, is the presence of extrafloral nectaries (EFNs). Extrafloral nectaries are sugar-producing glands (Bentley 1977). They occur on many plant organs such as leaf blades, petioles, stems or close to the reproductive organs and are found in a wide variety of plant taxa. Their adaptive significance has long been debated. In short, the 'protectionists' support the idea that ants visiting the EFNs protect this resource and hence the plant from being eaten by other herbivores (Protection or Ant-guard hypothesis) (Bentley 1976, Freitas and Oliveira 1996). The 'exploitationists' state that the nectar glands have a physiological function not relevant to any mutualism between ants and plants (O'Dowd and Catchpole 1983) and interactions with ants are a by-product of this physiological function.

There are some requirements if the beneficial interactions between ants and plants bearing EFNs are to work. Ants must be present on the plants and be aggressive towards potential herbivores and ideally eat them. The plants, in turn, must be vulnerable to herbivores at least during some stage of their life, which reduces their fitness. Ideally, nectar flow should vary with the life stage of the plant and occur when it is most needed, for example when most likely to be attacked by herbivores. There is considerable evidence that this is the case but the relative magnitude of benefits and costs is difficult to quantify. Strong support for the role of EFNs as an ant-mediated herbivore defence system comes from the observation that in habitats lacking ants, plants with EFNs are much rarer than those without EFNs. For example, nectary plants and ants are rare at high altitudes in the tropics (Keeler 1979), where the moist cool climate provides less favourable conditions for ground-dwelling ants, and in habitats which historically lack ants the percentage of plants with EFNs is low. Hawaii, which at no time was connected to a continent and has no native species of ant, has a flora that is poor in EFNs, which is consistent with the hypothesis that nectaries as part of an anti-herbivore defence system do not function in the absence of ants (Keeler 1985).

Now, considering the ant–plant defence system via EFNs and the ant–homopteran–lycaenid/nectar production system gives an interesting meta-mutualism configuration in which two separate mutualistic interactions meet and might change the strategies and affect the evolution of the partners of ants. A number of non-mutually exclusive hypotheses about the outcome of such interactions are proposed (Becerra and Venable 1989, 1991, Offenberg 2000) and three are explained here for an ant–aphid–EFN system (Fig. 6.28). These hypotheses basically view the mutualistic meta-community either from the plant or aphid perspectives or both.

(1) *Exploitation hypothesis*

Under this scenario aphids feed on plants with EFNs because they attract ants, which may also benefit the honeydew producers. Thus, aphids may actually

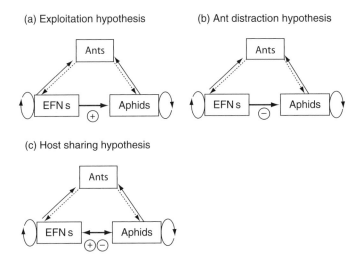

Fig. 6.28. Hypothetical interactions between extrafloral nectaries (EFNs) and honeydew/nectar producing partners of ants. + indicates positive effects, − negative effects of EFNs on ant partners. Circles indicate the costs of developing EFNs, producing honeydew, nectar or secretory glands that provide ants with sugary excreta. For example, in (a) producing EFNs or nectar is costly and the benefits of ant attendance is higher for aphids than for the plants bearing EFNs.

exploit the interaction between plants and ants, especially if they are able to reduce their costs of producing attractive or large quantities of honeydew. They can be called cheaters or parasites of the ant–plant mutualism. As a consequence, plants may experience more damage by herbivorous insects if ants provide less protection.

(2) *Ant distraction hypothesis*

Here the assumption is that while ants attend their honeydew and nectar producing partners, plants use EFNs to distract ants from their ant mutualists (Becerra and Venable 1989, Koptur 1991). In following this strategy, the benefit of reduced ant-dependent herbivory must be larger than the costs of producing nectar and the associated morphological structures (Keeler 1981). This implies that ant-dependent insects and plants compete for ants but it does not directly stress the protection provided by ants. Instead, it proposes that the main fitness benefit of EFNs is via interference; that is, a reduction in homopteran and lycaenid damage results from reduced protection by ants, which exposes them to a greater attack by natural enemies.

(3) *Host-sharing hypothesis*

Because similar selection pressures can operate as top-down forces on plants and herbivores, strategies to exploit the protection services of ants might result in the plants and their herbivores having a shared interest. For example, it is possible that unattended aphids are more likely to be preyed upon by ants on plants with EFNs (Engel *et al.* 2001). Thus, exploiting plants with EFNs would require the evolution of mechanisms that make their interactions with ants less damaging. For plants, in contrast, investment in EFNs only makes sense if there are ants in the habitat that can be used as a defence against herbivores. Plants with EFNs and honeydew/nectar producing insects may benefit if they are more attractive to ants and as a consequence of offering a continuously available resource the ants shift their nests to these more permanent resources. In this way the ants would be shared by the plants and homopterans they host provided the damage inflicted by the sap suckers on their host plant is smaller than that of non-ant-attended herbivores. From a plant's perspective it is possibly more difficult to defend against sap-sucking insects by means of EFNs because these potential mutualists of ants use the same resource as the plants, phloem sap, which cannot usually be loaded with secondary metabolites like other plant tissues.

To make the arguments more quantitative, a survey of 7 different families of aphids occurring in Denmark and Fennoscandia revealed that 14 ant-attended and 14 unattended species feed on plants with EFNs, while 84 attended and 335 unattended aphids feed on plants without EFNs. This suggests that overall aphids are positively associated with ants and with EFNs (Offenberg 2000). However, such species counts do not provide independent tests because the phylogeny of the aphids is often unknown below the family level.

Competition between plants with EFNs and honeydew/nectar producers, and between different species of myrmecophilous insects is a 'red queen game' as the need to produce more attractive honeydew than a potential competitor suggests that alternative resources should affect this mutualistic metacommunity, for example in situations where partners of ants feed on plants with EFNs or food bodies. However, the outcome of such interactions is equivocal, with some studies reporting no competition between EFNs and honeydew-producing Homoptera (Del-Claro and Oliveira 1993) and others reporting an effect (Koptur 1991, Sakata and Hashimoto 2000, Engel *et al.* 2001). For example, Blüthgen *et al.* (2000) studied the relationship between ants and plants and between ants and homopterans in a lowland Amazonian rainforest in Venezuela. In this study area sugar was abundantly available from either homopterans (Coccidae and Membracidae) feeding on trees or EFNs on lianas in these arboreal metamutualistic communities. The majority of the ant species collected in traps had fed on one or both of these sugar sources and their number was significantly higher on plants that offered such resources compared with those without ant partners or EFNs. In addition, the density of ant workers on plants with EFNs was similar to that on those with homopteran honeydew. This would appear to support the host-sharing hypothesis. However, a closer analysis at the species level suggests more complex community processes. Honeydew resources were monopolized by one or a few dominant aggressive ants, while on EFN plants a large diversity of subdominant ant species prevailed. In addition, preferences were recorded for dominant ants associated with particular species of homopterans (e.g. *Azteca, Dolichoderus, Pheidole*) and even with particular tree species. This suggests that different competitive abilities, resource preferences and availability of different resources result in ant mosaics at the local scale, which argues against the host-sharing hypothesis. Most likely density dependent effects also play a role but virtually nothing is known about how the relative quantities of different honeydew/nectar resources affect these mutualisms, especially in the tropics.

A more complete picture is presented by Engel *et al.* (2001) who describe a mutualistic metacommunity consisting of broad bean (*Vicia faba*), an ant (*Lasius niger*) and a facultative myrmecophile (*Aphis fabae*) and an unattended aphid species (*Acyrthosiphum pisum*). Small colonies of *A. fabae* did not distract ants from EFNs. This does not accord with interference competition, but supports host sharing. However, colonies of *A. pisum* declined on *V. faba* plants with active EFNs, suggesting some protection against sucking herbivores at least at some time in a season. *Vicia faba*, though, is not native to Europe, and in habitats with different ant species other outcomes are possible. Nevertheless, the message of this study is again, depending on the level of analysis, that different hypotheses regarding the function of EFNs within a sap-sucker metacommunity might be either accepted or rejected. In particular, it is necessary to record the density dependent costs and benefits in EFNs–ant–ant-partner systems. Currently, the different hypotheses are unable to discriminate between different mechanisms or cause and effect in these associations (Offenberg 2000).

IN SUMMARY, experimental evidence of multiple mutualistic interactions that involve more than pair-wise interactions between ants and their insect partners is growing and increasingly demonstrating the involvement of density dependence, environmental, bottom-up and top-down effects as well as competition-colonization trade-offs in the outcome of these interactions. These examples clearly show that a resource-based approach is beneficial for understanding the mechanisms that affect mutualism. It will remain a challenging task, however, to pinpoint the relative importance of the mechanisms involved in time and space.

6.5.4 General conclusions

For several decades ecologists have argued about the importance of the various processes that affect community structure and determine population sizes and stability. These arguments are based on mathematical models, observations and experiments. While these different approaches are clearly biased to antagonistic interactions, such as interspecific competition, predation or parasitism, they are increasingly being combined with environmental variability, time lags or spatial features. Only recently this list was expanded to include mutualism and hence more than two trophic levels. It is unlikely that a single factor will explain the patterns of association between ants and their partners, not even in a particular system. The reason is that mutualisms between insect species are inherently multitrophic, multidimensional and temporally variable associations. This raises an important question. Does the conclusion that no single factor can explain the organization of herbivorous insect–ant communities mean that no predictions or generalizations are possible? We believe that generalizations are possible, but they must be made conditional upon a variety of life-history, environmental and community characteristics (see, for example, Schoener 1986).

To understand the organization of an assemblage of mutualistic and non-mutualistic herbivores the details of the relationships of each of the individual herbivore species with its predators, competitors, mutualists, seasonal change in host plant quality, density changes and spatial distribution need to be considered. Thus, the limited energy and financial resources should be focused on a few systems, which can be reasonably studied experimentally at different spatial and temporal scales. These systems should be complementary in nature and stimulate theoretical and experimental approaches. Ants and their insect partners certainly qualify because detailed knowledge is available on many population and community aspects, such as density dependent processes, competition and predation and habitat fragmentation. In this respect the argument that aphids, coccids or membracids are special cases and unsuitable for drawing meaningful and general conclusions misses the point. If community processes are embedded in a hierarchy of processes then the performance of cross-scale studies should allow general predictions like those depicted in Fig. 6.29.

For example, it is reasonable to assume that if predation or parasitism of the partners of ants is intense, host plant quality and interspecific/intraspecific competition are less likely to affect the dynamics of the herbivore; that is, plant quality becomes relatively unimportant for the herbivore. Under these conditions, mutualistic relationships (and the exploitation of mutualism), however, are likely to gain in relative importance. On the other hand, if natural enemies play a negligible role in the population dynamics of the partners of ants, then bottom-up effects gain in relative importance. Under these conditions host quality and competitive effects also gain in relative importance. For example,

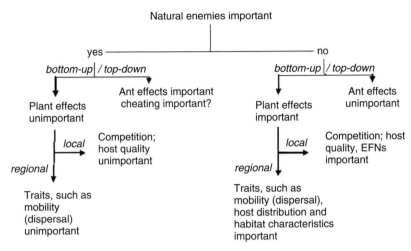

Fig. 6.29. Simplified hierarchical view of the relative importance of different traits affecting populations and communities, in particular, the impact of natural enemies on mutualistic associations between ants and their partners. This dichotomy is based on the assumption that the presence/absence of natural enemies is a major driving factor in the development of mutualisms between ants and their insect partners. The competition–colonization trade-off, bottom-up and top-down processes at local and regional scales are integral parts mediating between different spatial scales. Seasonal changes in these traits are not represented but most likely affect the relative importance of the traits.

Edson (1985) showed that competition between goldenrod aphids was reduced when they were able to move between goldenrod plants. Of course the relative importance of these effects could change in different years giving the above dichotomy of processes different weights at different times. It is also conceivable that specialists with a narrow diet breadth are more likely to be affected by the attributes of their host (i.e. quality) than generalist herbivorous partners of ants. Obligate myrmecophiles will almost always be relatively more affected by selection pressures that operate on ants and relatively less so by plant traits, than facultative myrmecophiles. The trade-off between competition and colonization in myrmecophile and non-myrmecophile ant partners is likely to be important in any of the above scenarios, but probably more so when natural enemies significantly affect the local abundance of the species.

The conditions mentioned above acknowledge the varying importance of plant quality, competition and mutualist traits, but are not sufficient to describe these dynamic reciprocal interactions. It is important to understand that attempts to find generalities must consider the conditions. Earlier and still rather influential studies hypothesized that knowledge of the trophic position of an organism is sufficient to provide a good indicator of the importance of competition and predation acting on an organism (Hairston *et al.* 1960). As shown in this chapter, there have been recent attempts to integrate more biological details and larger spatial and temporal scales when determining the relative strengths of host plant traits, life-history features of the mutualistic partners, dispersal ability, competition with other mutualists or EFNs, density dependent costs and benefits, predation and disturbance in a fragmented landscape, and predicting the outcome in terms of the mutualism–antagonism continuum. Collecting sufficient information to meet these requirements is not easy, but because mutualism touches upon each of these important ecological issues it is likely that cross-scale studies on mutualisms between ants and their insect partners will have a lasting effect on modern ecology.

7

Prospects and conclusions

Historically, ecology has been dominated by research on negative (antagonistic) interspecific interactions like competition and predation/parasitism. Nevertheless, this bias has led to a good understanding of the population dynamics of many species and organization of communities. In particular, the development of the theory of host–parasitoid and predator–prey dynamics was associated with experimental studies searching for mechanisms and general rules in population ecology. The advantage of studying negative interactions is that the fitness of the victim is likely to be zero, or close to zero, if eaten or parasitized. This makes it easier to track the outcome of these interactions both theoretically and experimentally. In addition, the idea of regulation is very seductive as it suggests clear relationships between predators and prey or between hosts and parasites and a simple mechanism generating population fluctuations. Mutualism, in contrast, involves reciprocal positive interactions between organisms belonging to different species and often produces less clear cut outcomes. Positive interactions tend to be diffuse, dependent on boundary conditions and thus may shift from positive to negative over time. This means that a good understanding of the conditions is necessary for quantifying the net outcome of conditional interactions. In spite of these difficulties, there is growing evidence that at least temporal positive interactions are widespread in insect communities and, in particular, between ants and their insect partners. It is argued here that mutualism, similarly to competition/predation, is an important ecological force determining individual fitness, population dynamics and community structure of dominant insects, such as ants, aphids, coccids and membracids. Studies on mutualism are therefore likely to make a significant contribution to modern ecology. Below are summarized the most important levels of organization at which mutualisms can be studied, because mutualism between ants and their partners is first and foremost a multi-level issue.

7.1 Life-history level

There are many case studies that report positive, negative or neutral effects of ants on the fitness of other insects. Less clear is the relative fitness gain to the attending ants, especially if conditions vary. It is likely that the costs and benefits are asymmetric and dependent on the life histories of the partners. For example, ants and partners of ants have usually short and often different generation times. In particular, generation times of parthenogenetically reproducing aphids are significantly shorter than those of coccids or lycaenids. However, it is difficult to predict how asymmetric generation times affect the outcome of mutualism. The generation time ratio (GTR) hypothesis might be useful for this. Provided that the GTR hypothesis applies to the negative interactions between aphids and their predators (D(predators)/D(prey) $=$ GTR), with GTR > 1 suggesting no significant impact of natural enemies on their prey, the same principle could be applied to mutualistic associations between ants and their partners (positive interactions). Large generation time ratios of ants and their partners (D(ants)/D(partners of ants)) > 1 might suggest a less close association because the long-lived partner (ants) is unlikely to respond quickly to the changing densities of the short-lived partner. For example, there is no evidence that ants actually regulate the density of a colony of honeydew producers according to their needs. Rather they abandon whole colonies and attend those closer to their nest. In addition, short generation times and high growth rates make the protection function of ants less important because losses due to natural enemies become quantitatively insignificant. As aphids have the shortest generation times of all insect partners of ants their association with ants should be more facultative, opportunistic, asymmetric, condition dependent and probably evolutionarily more labile than that between ants and coccids, lycaenids or membracids. As shown in previous chapters preliminary results support such a prediction.

Of course, GTRs are only part of the multifactorial interaction between ants and their partners: because these interactions are embedded in spatial settings and bottom-up/top-down forces they make the relative competition–colonization abilities of both partners important features, which ultimately determine the degree of interaction along the mutualism–antagonism continuum. What is needed is more information at the level of individual mutualists, for example quantification of the fitness costs and benefits of socially organized ants and their partners with different modes of reproduction. Sorely needed is more information on the response of each partner under varying environmental conditions. This information will help to identify patterns and determine the relative importance of the different factors affecting the outcome of these interactions.

7.2 Population level

At the population level it is even more difficult to measure costs and benefits in the mutualisms between ants and their partners. This is because they are embedded in top-down and bottom-up processes. Both ants and partners of ants are vulnerable to competition, predation, parasitism and cheating, and depend, for example, on their host plants and their spatial configuration. As a consequence, changes other than those associated with the immediate pairwise interactions are likely to affect the outcome of mutualisms. Due to these unspecific multilevel effects, the presence of significant time lags in population growth and the local nature of interactions between individuals of different populations, mutualisms are often termed diffuse. However, a more adequate description of the strength of these interactions is that they are condition dependent and it is the aim of science to identify the conditions.

One condition is the density of the partners that interact. More recent empirical and theoretical studies emphasize density as a key parameter, which may cause a shift from a mutualistic to an antagonistic interaction, or vice versa (Neuhauser and Fargione 2004). If one resource (partner) is abundant then it is a commodity of low value and from the point of view of the abundant resource the benefits of interacting with a rare partner become less important. Indeed, many empirical and theoretical studies suggest that abundance results in antagonism, while rarity increases the value of a partner and thus the benefits that can be derived by positively interacting with it. However, at low densities of partners, or when they are rare, mutualism might not work because the pay-offs from investing in mutualism might be too uncertain and too variable.

In recent years population and community ecologists have increasingly focused on the spatial context in which populations interact (Hanski and Gilpin 1997). This has stimulated theoretical studies on mutualistic interactions, but progress is hampered by lack of empirical data and tests. For example, it is often unclear whether effects of ants described at the life-history level have any effects at larger scales. For example, it is often unknown whether the population dynamics of attended and unattended species differ. In particular, it is conceivable that apparent benefits at the individual or colony level might negatively affect the overall population of a partner. Especially, when local competition and colonization of new hosts are alternative strategies linked via trade-offs, interference of ants with these attributes might pose problems to partners of ants. For example, it could be disadvantageous if attendance leads to costs for a facultative partner, which as a consequence shows lower growth rates or delayed dispersal.

Thompson (1994) strongly emphasized the general importance of spatial heterogeneity for coevolution. Although most well-studied examples of mutualism involve plant–insect interactions, the basic idea is equally applicable to ants and their partners. Local populations differ in the traits shaped by an interaction because conditions vary locally. As a consequence, traits of local populations might be well matched for mutualisms in some patches but not in others. Most likely, few traits will be globally advantageous when interacting with ants. For example, the ability to colonize new plants is unimportant if one is feeding on high-quality plants or if the population size of the colony is relatively small and competition is weak. In this case being a good disperser might be maladaptive. In contrast, when the local quality of a patch is deteriorating or when there is intense competition, those individuals with good dispersal abilities might be fitter. Interactions with ants, however, might be negatively linked to fitness if dispersal becomes important, because ants tend to prey on moving objects. All insect partners of ants are phytophagous and therefore face these trade-offs and the associated costs and benefits.

Some open questions result from this mosaic view of interactions, including both bottom-up and top-down aspects and the associated colonization–competition dichotomy. How often are mutualistic populations limited by access to resources rather than to mutualistic partners, and what are the relative costs and benefits of these limitations? Is there a general relationship between competition–colonization ability and the strength of mutualistic interactions? Should species with low colonization abilities tend to be more mutualistic? Is there an optimal colony size at which the benefits of mutualistic relationships are maximal? Can one species of aphid, coccid, lycaenid, etc. really negatively affect mutualistic relationships between ants and another honeydew/nectar producer? Is the risk of exploitation of mutualistic relationships between ants and their insect partners important and does this risk affect colonization–competition trade-offs? How important is intraguild competition for the services of ants among mutualists with different life histories (GTRs)? We suspect that in those cases in which GTRs between exploiters and mutualists are small exploitation risks are also small.

In the immediate future there are a number of interesting questions that should be addressed, either experimentally or theoretically. For example, do the population dynamics of myrmecophiles and non-myrmecophiles differ, and if so, in what way? How do costs and benefits vary with colony size and during the course of a season? How can costs and benefits be determined considering the different genetic structures of the clonal and socially organized partners? Are there differences in the dispersal rates/patterns of myrmecophiles and non-myrmecophiles? How does patch size affect ant attendance and what are the costs for ants in attending different partners? How does the spatial and temporal variability in plant/patch phenology/quality affect the distribution and abundance of ant partners showing different degrees of associations with ants? What is the relative importance, in a temporal and spatial context, of the factors depicted in Fig. 5.1 for the outcome of mutualistic relationships?

However, given the conditionality of the interactions, which depend, for example, on different life histories, genetic and social organization or environmental configurations, the expectation is that obligate mutualisms between ants and their partners should be rare.

7.3 Community level

There is growing interest in the role of mutualism in structuring communities, both in aquatic and terrestrial systems (Bronstein 1994b, Hay *et al.* 2004). Generally, communities are organized via trophic interactions and mutualisms should thus be viewed in a food web (resource based) context rather than a pair-wise interaction of partners.

The community context of mutualisms between ants and their insect partners is presented in Fig. 6.20. However, it is notoriously difficult to demonstrate experimentally the relative importance of the effects of different factors on mutualism. For example, it would be useful to have a rough idea of how fitness costs and benefits for mutualists vary with resource availability and the density of the partners of ants and natural enemies or cheaters. Field surveys addressing these questions are virtually non-existent. However, drawing conclusions about the importance of mutualistic interactions for ecological communities depends on identifying a fitness currency comparable to predation or parasitism, so that immediate consequences of interactions are discernible. The continuing controversy about the relative importance of top-down and bottom-up forces, for example, in the population dynamics of aphids, might well be resolved if one considers the spatial aspects more closely. At the local scale predators might eliminate a population of aphids or membracids, but at the regional scale their influence might be negligible. As a consequence, in small-scale studies it is often argued that the protection against natural enemies afforded by ants to their partners is of paramount importance. However, while local populations of these partners of ants might be eliminated, repeated colonization of empty patches combined with high rates of reproduction might make this mortality factor less important. Again, this argues for an integrated approach incorporating the simultaneous study of multiple levels of interactions when trying to understand the outcome and evolution of positive and negative interactions between ants and their partners.

The effects of mutualistic interactions on communities are easier to assess, for example on the diversity of those species not involved in mutualism. Mutualisms between ants and their partners are likely to have a strong effect on community structure because ants are an ecologically dominant group of insects. Recent studies show convincingly that the interactions between ants and aphids, ants and membracids, etc. significantly reduce insect species richness and abundance of other insects on their host plants (Wimp and Whitham 2001, Kaplan and Eubanks 2002, Renault *et al.* 2005). These effects are direct, via the attraction of tending ants, which remove other herbivores, or indirect via their growing population size and feeding activities ultimately outcompeting other non-myrmecophile herbivores. Interactions between aphids and dominant ants also affect the local ant community, which suggests that these mutualisms, no matter how temporal and local they may be, are a key interaction and have community-wide effects. A surprising result, however, is that in spite of the importance of mutualism it does not appear to be a radiation platform for species diversification; that is, the ability to associate with ants does not appear to boost speciation (see, for example, Section 4.1.5). The most likely explanation for this is that positive interactions, although prevailing, are highly dependent on local conditions and potential benefits are constrained by bottom-up and top-down processes. Therefore, the ability to associate with ants remains a labile trait. There are probably no habitats for ants and their potential partners that are sufficiently adverse that it is only by co-operating that each partner can increase its fitness.

Finally, we highlight some unresolved questions that need to be addressed in future studies on the role of positive and negative interactions between ants and their partners in structuring communities. For example, how do multiple species coexist within guilds of mutualists? How do mutualists persist in the presence of antagonists/cheaters/predators? Are mutualistic relationships in insects more likely between organisms belonging to different trophic levels/ particular trophic communities? Is mutualism more likely between insects whose population organization shows high intra-deme relationships? In what way does the spatial or temporal context affect the strength of competitive and mutualistic interactions? How much asymmetry in the costs and benefits for both partners can be tolerated in mutualistic communities before such a relationship collapses into an antagonistic one? What is the relative role of mutualistic and antagonistic relationships in community organization? Are there general patterns in the life histories of mutualists or are spatial attributes of patches important for understanding the relative role of these factors in determining the outcome of the interactions? Put more directly: how important are the biological details relative to the spatial configuration of interacting populations in maintaining mutualistic relationships?

7.4 Ecosystem level

The importance of mutualistic interactions at local and regional scales extends to the level of ecosystems. In particular, ecosystem consequences of ant–myrmecophile interactions have been shown for invasive ants like *Solenopsis invicta* and for invasive honeydew producers, which displace the native communities of sap feeders (Gotelli and Arnett 2000). The success of invasive species often rests on their variable nest densities and social structures (Ingram 2002a, b, Morrison 2002, Tsutsui and Suarez 2003). In addition, establishing new trophic relationships is likely to promote the success of invaders. For example, the invasive mealybug (*Antonia graminis*) provides about half of the daily energy requirement of *Solenopsis invicta*, an invasive ant in Texas (Helms and Vinson 2002). McGlynn (1999) lists about 150 ant species that are recorded outside their native habitat. Most of them are cryptic opportunists that attend honeydew producers. The ability of these invasive ants of forming new associations with honeydew producers is suggested to be the main factor determining their success and potentially that of the sap feeders they associate with (Abbott and Green 2007). A negative impact of ants and homopterans on the native fauna may be offset by short-term positive changes, for example a decrease in more serious plant pests. This was why invasive ant species, like the Argentine ant (*Linepithema humile*), were introduced to control herbivorous insects in agricultural areas (Holway *et al.* 2002a).

   In addition to the invasive species, ants and many homopterans are dominant insect groups in many different types of habitats and affect large-scale ecosystem processes. For example, by preying on other arthropods or tending honeydew producers and by large-scale burrowing activities, they affect nutrient fluxes through ecosystems and decomposition processes (Stadler *et al.* 2006b). Similarly, an increase in honeydew availability due to increase in population sizes of attended species can significantly increase organic and inorganic carbon and nitrogen fluxes through ecosystems (Stadler *et al.* 1998, 2006a) and the increase in population size may significantly affect nutrient availability, microbial growth and decomposition. These effects of mutualists are underappreciated because multidisciplinary approaches are required to study the large-scale outcome of these interactions. Nevertheless, there is considerable evidence that mutualistic interactions between ants and their partners extend to the ecosystem level. These large-scale effects on ecosystem processes are most easily recognized when a species is introduced because it is initially not integrated into a complex food web. Thus their effects on ecosystems are often more easily identifiable compared with those of indigenous insects, which are woven into a tight net of interactions.

7.5 Conclusions

The study of mutualistic interactions between ants and their insect partners is often difficult because of the many factors that simultaneously influence the outcome of these relationships. For example, bottom-up effects like the distribution and abundance of the host plants of the phytophages or community composition of soil invertebrates are likely to cascade through the system and affect the growth rates of aphids, coccids, membracids and lycaenids either directly or indirectly via plant nutrients or secondary plant compounds. In a similar way, top-down effects of natural enemies are likely to shape the strength of ant–myrmecophile relationships. As a consequence of these external forces, a high degree of specialization in these interactions is not expected and most mutual associations are best characterized as temporary and opportunistic. This is probably true of a large number of the relationships between ants and their partners. For example, even though there are spectacular examples of highly specialized associations between ants and aphids, coccids, membracids or lycaenids the majority of associations are facultative and unspecific (Bristow 1991, Fiedler 1991, Pierce *et al.* 2002, Stadler *et al.* 2003, Stadler and Dixon 2005) with an overall decline in numbers of myrmecophilous species from the tropics to the temperate regions (Table 4.1) (Pierce *et al.* 2002).

Given their multi-trophic nature the outcome of these relationships needs to be followed for a complete season or several seasons in order to appreciate the full range of costs and benefits for both partners. Laboratory experiments performed under constant conditions can only provide an incomplete understanding of the constraints acting on both partners and need to be paralleled by field investigations whose duration is sufficiently long to encompass the life cycles of both partners. It is important to determine fitness costs and benefits in different environments. Although it is experimentally challenging to simultaneously determine fitness parameters, such as intrinsic rates of population increase, it is more instructive to compare and understand these fitness costs and benefits for different species in different environments.

Our current understanding of fitness costs is still surprisingly rudimentary. For example, at the physiological level there is little or no information on the changing composition of honeydew/nectar and associated adaptation costs for different generations of aphids, coccids or lycaenids in response to ant attendance, in particular, when both colony size and quality of the host plants vary. Clearly, there is a need to determine the extent to which foraging activity is associated with sugar:protein imbalances in the diet of ants, and their effect on the communities of honeydew producers. In terms of ecology there is a growing awareness that spatial variability affects the distribution and abundance of both partners and the effects of bottom-up and top-down forces (Edson 1985, Müller and Godfray 1999, Gotelli and Ellison 2002, Blossey and Hunt-Joshi 2003). This indicates that a more spatially explicit or meta-community perspective of mutualism may be more appropriate (Addicott 1978b, Edson 1985, Albrecht and Gotelli 2001, Mooney and Tillberg 2005). The relative importance of mutualistic, neutral and antagonistic interactions between ants and their partners, and their relative role in community structure and species diversity are beginning to be addressed (Wimp and Whitham 2001). Still further in the future but no less important is the need to address the effects of the mutualistic and antagonistic interactions of these dominant groups of insects on nutrient cycling and ecosystem functioning (Loreau 1995, Stadler *et al.* 1998).

The traditional view is that ants are in control of the interaction with their partners. However, many species of honeydew producers do not compete for the services of ants and appear to have a range of options to cope with ants that are unpredictable and unreliable mutualists. Considering the different life cycles (e.g. parthenogenetic reproduction versus social organization) and the different selection pressures that are associated with these features, there is ample opportunity for both partners to exploit each other. For example, for a clonal fast-reproducing organism, there might be little cost involved if ants prey on a few individuals. However, active foragers need to monopolize a source of energy within their foraging area in order to avoid conflicts with conspecifics or other ant species.

IN SUMMARY, recent advances in modelling mutualistic interactions incorporating life-history features and density dependent changes in the outcome of the interactions provide a starting point for exploring the whole range of interactions and condition dependent costs and benefits. On the experimental side, the relationships between ants and their partners are easy to manipulate both in the field and laboratory. Based on a good knowledge of the life histories and population biology of many species these associations appear to be ideal systems for studying the driving forces in the ecology and evolution of antagonistic/mutualistic relationships. Despite Janzen's (1985) claim, mutualism does not appear to have been thought to death. Adopting a less myrmecologist-centred point of view reveals that there are many unresolved and challenging questions and the study of mutualisms between ants and their partners offers excellent opportunities for resolving these questions. Identifying the mechanisms that drive the evolution of mutualism as well as finding the patterns that constrain or favour positive interactions between ants and their insect partners will prove a rewarding task for many years to come.

References

Abbott, K. I., and P. Green. 2007. Collapse of an ant–scale mutualism in a rainforest on Christmas Island. *Oikos* **116**:1238–1246.

Abrams, P. A., and H. Matsuda. 1996. Positive indirect effects between prey species that share predators. *Ecology* **77**:610–616.

Abrams, P. A., and W. G. Wilson. 2004. Coexistence of competitors in metacommunities due to spatial variation in resource growth rates; does R* predict the outcome of competition? *Ecology Letters* **7**:929–940.

Abrams, P. A., R. D. Holt, and J. D. Roth. 1998. Apparent competition or apparent mutualism? Shared predation when populations cycle. *Ecology* **79**:201–212.

Addicott, J. F. 1978a. Competition for mutualists – aphids and ants. *Canadian Journal of Zoology* **56**:2093–2096.

Addicott, J. F. 1978b. The population dynamics of aphids on fireweed: a comparison of local populations and metapopulations. *Canadian Journal of Zoology* **56**:2554–2564.

Addicott, J. F. 1979. A multispecies aphid-ant association: density dependence and species-specific effects. *Canadian Journal of Zoology* **57**:558–569.

Addicott, J. F. 1981. Stability properties of 2-species models of mutualism: simulation studies. *Oecologia* **49**:42–49.

Addicott, J. F., J. M. Aho, M. F. Antolin, D. K. Padilla, J. S. Richardson, and D. A. Soluk. 1987. Ecological neighborhoods: scaling environmental patterns. *Oikos* **49**:340–346.

Agrawal, A. A., and J. A. Fordyce. 2000. Induced indirect defence in a lycaenid-ant association: the regulation of a resource in a mutualism. *Proceedings of the Royal Society of London Series B* **267**:1857–1861.

Agrawal, A. A., N. Underwood, and J. R. Stinchcombe. 2004. Intraspecific variation in the strength of density dependence in aphid populations. *Ecological Entomology* **29**:521–526.

Albrecht, M., and N. J. Gotelli. 2001. Spatial and temporal niche partitioning in grassland ants. *Oecologia* **126**:134–141.

Allee, W. C. 1938. *The Social Life of Animals*. Boston: Beacon Press.

Allee, W. C. 1949. *Principles of Animal Ecology*. Philadelphia: W. B. Saunders Co.

Amarasekare, P. 1998a. Allee effects in metapopulation dynamics. *American Naturalist* **152**:298–302.

Amarasekare, P. 1998b. Interactions between local dynamics and dispersal: insights from single species models. *Theoretical Population Biology* **53**:44–59.

Amarasekare, P. 2004. Spatial dynamics of mutualistic interactions. *Journal of Animal Ecology* **73**:128–142.

Andersen, M. 1991. An ant-aphid interaction: *Formica fusca* and *Aphthargelia symphoricarpi* on Mount St. Helens. *American Midland Naturalist* **125**:29–36.

Antolin, M. F., and J. F. Addicott. 1991. Colonization, among shoot movement, and local population neighborhoods of two aphid species. *Oikos* **61**:45–53.

Aoki, S. 1978. Two pemphigids with first instar larvae attacking predatory intruders (Homoptera, Aphidoidea). *New Entomologist* **27**:7–12.

Ashford, D. A., W. A. Smith, and A. E. Douglas. 2000. Living on a high sugar diet: the fate of sucrose ingested by a phloem-feeding insect, the pea aphid *Acyrthosiphon pisum. Journal of Insect Physiology* **46**:335–341.

Atsatt, P. R. 1981. Lycaenid butterflies and ants: selection for enemy free space. *American Naturalist* **118**:638–654.

Aviles, L. 1999. Cooperation and non-linear dynamics: an ecological perspective on the evolution of sociality. *Evolutionary Ecology Research* **1**:459–477.

Awmack, C. S., and S. R. Leather. 2002. Host plant quality and fecundity in herbivorous insects. *Annual Review of Entomology* **47**:817–844.

Axelrod, R., and W. D. Hamilton. 1981. The evolution of cooperation. *Science* **211**:1390–1396.

Axen, A. H. 2000. Variation in behavior of lycaenid larvae when attended by different ant species. *Evolutionary Ecology* **14**:611–625.

Axen, A. H., and N. E. Pierce. 1998. Aggregation as a cost-reducing strategy for lycaenid larvae. *Behavioral Ecology* **9**:109–115.

Axen, A. H., O. Leimar, and V. Hoffman. 1996. Signalling in a mutualistic interaction. *Animal Behaviour* **52**:321–333.

Azcarate, F. M., L. Arqueros, A. M. Sanchez, and B. Peco. 2005. Seed and fruit selection by harvester ants, *Messor barbarus*, in Mediterranean grassland and scrubland. *Functional Ecology* **19**:273–283.

Bach, C. E. 1991. Direct and indirect interactions between ants (*Pheidole megacephala*), scales (*Coccus viridis*) and plants (*Pluches indica*). *Oecologia* **87**:233–239.

Bartlett, B. R. 1961. The influence of ants upon parasites, predators, and scale insects. *Annals of the Entomological Society of America* **54**:543–551.

Baylis, M., and N. E. Pierce. 1991. The effect of host plant quality on the survival of larvae and oviposition by adults of an ant-tended lycaenid butterfly, *Jalmenus evagoras. Ecological Entomology* **16**:1–9.

Beattie, A. J. 1976. Plant dispersion, pollination and gene flow in *Viola. Oecologica* **25**:291–300.

Beattie, A. J. 1985. *The Evolutionary Ecology of Ant-Plant Mutualisms*. Cambridge: Cambridge University Press.

Becerra, J. X. I., and D. L. Venable. 1989. Extrafloral nectaries: a defense against ant-homoptera mutualisms? *Oikos* **55**:276–280.

Becerra, J. X., and D. L. Venable. 1991. The role of ant-Homoptera mutualisms in the evolution of extrafloral nectaries. *Oikos* **60**:105–106.

Beckerman, A., T. G. Benton, E. Ranta, V. Kaitala, and P. Lundberg. 2002. Population dynamic consequences of delayed life-history effects. *Trends in Ecology and Evolution* **17**:263–269.

Bell, G. 2001. Neutral macroecology. *Science* **293**:2413–2418.

Bentley, B. L. 1976. Plants bearing extrafloral nectaries and the associated ant community: interhabitat differences in the reduction of herbivore damage. *Ecology* **57**:815–820.

Bentley, B. L. 1977. Extrafloral nectaries and protection by pugnacious bodyguards. *Annual Review of Ecology and Systematics* **8**:407–427.

Bergeson, E., and F. J. Messina. 1997. Resource- versus enemy-mediated interactions between cereal aphids (Homoptera: Aphididae) on a common host plant. *Annals of the Entomological Society of America* **90**:425–432.

Bergeson, E., and F. J. Messina. 1998. Effect of a co-occurring aphid on the susceptibility of the Russian wheat aphid to lacewing predators. *Entomologia Experimentalis et Applicata* **87**:103–108.

Berryman, A. A., M. Lima, and B. A. Hawkins. 2002. Population regulation, emergent properties, and a requiem for density dependence. *Oikos* **100**:600–636.

Bhatkar A. P., and W. J. Kloft. 1977. Evidence, using radioactive phosphorus, of interspecific food exchange in ants. *Nature* **265**: 140–142.

Billick, I., and K. Tonkel. 2003. The relative importance of spatial vs. temporal variability in generating a conditional mutualism. *Ecology* **84**:289–295.

Billick, I., M. Weidmann, and J. Reithel. 2001. The relationship between ant-tending and maternal care in the treehopper *Publilia modesta*. *Behavioral Ecology and Sociobiology* **51**:41–46.

Bishop, D. B., and C. M. Bristow. 2001. Effect of Allegheny mound ant (Hymenoptera : Formicidae) presence on homopteran and predator populations in Michigan jack pine forests. *Annals of the Entomological Society of America* **94**:33–40.

Blackman, R. L. and V. F. Eastop. 1994. *Aphids on the World's Trees*. Wallingford: CAB International.

Blossey, B., and T. R. Hunt-Joshi. 2003. Belowground herbivory by insects: influence on plants and aboveground herbivores. *Annual Review of Entomology* **48**:521–547.

Blüthgen, N., and K. Fiedler. 2004a. Competition for composition: lessons from nectar-feeding ant communities. *Ecology* **85**:1479–1485.

Blüthgen, N., and K. Fiedler. 2004b. Preferences for sugars and amino acids and their conditionality in a diverse nectar-feeding ant community. *Journal of Animal Ecology* **73**:155–166.

Blüthgen, N., G. Gebauer, and K. Fiedler. 2003. Disentangling a rainforest food web using stable isotopes: dietary diversity in a species-rich ant community. *Oecologia* **137**:426–435.

Blüthgen, N., M. Verhaagh, W. Goitia, K. Jaffe, W. Morawetz, and W. Barthlott. 2000. How plants shape the ant community in the Amazonian rainforest canopy: the key role of extrafloral nectaries and homopteran honeydew. *Oecologia* **125**:229–240.

Bolton, B. 1995. *A New General Catalogue of the Ants of the World*. Cambridge, MA: Harvard University Press.

Bonabeau, E., G. Theraulaz, and J. L. Deneubourg. 1999. Dominance orders in animal societies: the self-organization hypothesis revisited. *Bulletin of Mathematical Biology* **61**:727–757.

Bonabeau, E., G. Theraulaz, J. L. Deneubourg, S. Aron, and S. Camazine. 1997. Self-organization in social insects. *Trends in Ecology and Evolution* **12**:188–193.

Bonkowski, M., I. E. Geoghegan, A. N. E. Birch, and B. S. Griffiths. 2001. Effects of soil decomposer invertebrates (protozoa and earthworms) on an above-ground phytophagous insect (cereal aphid) mediated through changes in the host plant. *Oikos* **95**:441–450.

Bonsall, M. B., and M. P. Hassell. 1997. Apparent competition structures ecological assemblages. *Nature* **388**:371–373.

Bonsall, M. B., V. A. A. Jansen, and M. P. Hassell. 2004. Life history trade-offs assemble ecological guilds. *Science* **306**:111–114.

Boucher, D. H. 1985. *The Biology of Mutualism*. New York: Oxford University Press.

Boursaux-Eude, C., and R. Gross. 2000. New insights into symbiotic associations between ants and bacteria. *Research in Microbiology* **151**:513–519.

Bradley, G. A., and J. D. Hinks. 1968. Ants aphids and jack pine in Manitoba. *Canadian Entomologist* **100**:40–50.

Braschler, B., and B. Baur. 2003. Effects of experimental small-scale grassland fragmentation on spatial distribution, density, and persistence of ant nests. *Ecological Entomology* **28**:651–658.

Braschler, B., and B. Baur. 2005. Experimental small-scale grassland fragmentation alters competitive interactions among ant species. *Oecologia* **143**:291–300.

Braschler, B., G. Lampel, and B. Baur. 2003. Experimental small-scale grassland fragmentation alters aphid population dynamics. *Oikos* **100**:581–591.

Braschler, B., S. Zschokke, C. Dolt, G. H. Thommen, P. Oggier, and B. Baur. 2004. Grain-dependent relationships between plant productivity and invertebrate species richness and biomass in calcareous grasslands. *Basic and Applied Ecology* **5**:15–24.

Brauchli, K., T. Killingback, and M. Doebeli. 1999. Evolution of cooperation in spatially structured populations. *Journal of Theoretical Biology* **200**:405–417.

Breton, L. M., and J. F. Addicott. 1992a. Density-dependent mutualism in an aphid-ant interaction. *Ecology* **73**:2175–2180.

Breton, L. M., and J. F. Addicott. 1992b. Does host plant quality mediate aphid-ant mutualism? *Oikos* **63**:253–259.

Bristow, C. M. 1984. Differential benefits from ant-attendance to two species of Homoptera on New York ironweed. *Journal of Animal Ecology* **53**:715–726.

Bristow, C. M. 1991. Why are so few aphids ant-tended? In *Ant-plant Interactions*, ed. C. R. Huxley and D. F. Cutler, pp. 104–119. Oxford: Oxford University Press.

Brodbeck, B. V., and D. Strong. 1987. Amino acid nutrition of herbivorous insects and stress to host plants. In *Insect Outbreaks: Ecological and Evolutionary Perspectives*, ed. P. Barbosa and J. Schultz, pp. 347–364. New York: Academic Press.

Bronstein, J. L. 1988. Mutualism, antagonism, and the fig-pollinator interaction. *Ecology* **69**:1298–1302.

Bronstein, J. L. 1994a. Conditional outcomes in mutualistic interactions. *Trends in Ecology and Evolution* **9**:214–217.

Bronstein, J. L. 1994b. Our current understanding of mutualism. *Quarterly Review of Biology* **69**:31–51.

Bronstein, J. L. 2001. The exploitation of mutualism. *Ecology Letters* **4**:277–287.

Brown, J. L. 1983. Cooperation: a biologist's dilemma. *Advances in the Study of Behaviour* **13**:1–37.

Brown, M. J. F., and D. M. Gordon. 2000. How resources and encounters affect the distribution of foraging activity in a seed-harvesting ant. *Behavioral Ecology and Sociobiology* **47**:195–203.

Buckley, R. 1987. Ant-plant-homopteran interactions. *Advances in Ecological Research* **16**:53–85.

Buckley, R., and P. Gullan. 1991. More aggressive ant species (Hymenoptera, Formicidae) provide better protection for soft scales and mealybugs (Homoptera, Coccidae, Pseudococcidae). *Biotropica* **23**:282–286.

Cappuccino, N. 1987. Comparative population dynamics of two goldenrod aphids: spatial patterns and temporal constancy. *Ecology* **68**:1634–1646.

Cappuccino, N. 1988. Spatial patterns of goldenrod aphids and the response of enemies to patch density. *Oecologia* **76**:607–610.

Carroll, C. R., and D. H. Janzen. 1973. Ecology of foraging by ants. *Annual Review of Ecology and Systematics* **4**:231–257.

Choe, D.-H. and M. K. Rust. 2006. Ants learn the association between homopteran cuticular chemistry and honeydew. *Chemoecology* **16**: 175–178.

Cocroft, R. B. 1996. Insect vibrational defence signals. *Nature* **382**:679–680.

Cocroft, R. B. 1999. Parent-offspring communication in response to predators in a subsocial treehopper (Hemiptera : Membracidae : *Umbonia crassicornis*). *Ethology* **105**:553–568.

Cocroft, R. B. 2002. Antipredator defense as a limited resource: unequal predation risk in broods of an insect with maternal care. *Behavioral Ecology* **13**:125–133.

Cocroft, R. B., and R. L. Rodriguez. 2005. The behavioral ecology of insect vibrational communication. *BioScience* **55**:323–334.

Collins, C. M., and S. R. Leather. 2002. Ant-mediated dispersal of the black willow aphid *Pterocomma salicis* L.; does the ant *Lasius niger* L. judge aphid-host quality? *Ecological Entomology* **27**:238–241.

Connor, R. C. 1986. Pseudo reciprocity: investing in mutualism. *Animal Behaviour* **34**:1562–1566.

Connor, R. C. 1995. The benefits of mutualism: a conceptual framework. *Biological Review* **70**:427–457.

Costa, J. T., J. H. McDonald, and N. E. Pierce. 1996. The effect of ant association on the population genetics of the Australian butterfly, *Jalmenus evagoras* (Lepidoptera: Lycaenidae). *Biological Journal of the Linnean Society* **58**:287–306.

Cushman, J. H. 1991. Host-plant mediation of insect mutualisms: variable outcomes in herbivore-ant interactions. *Oikos* **61**:138–144.

Cushman, J. H., and J. F. Addicott. 1989. Intra- and interspecific competition for mutualists: ants as a limited and limiting resource for aphids. *Oecologia* **79**:315–321.

Cushman, J. H., and T. G. Whitham. 1989. Conditional mutualism in a membracid-ant association: temporal, age specific, and density dependent effects. *Ecology*, **70**:1040–1047.

Cushman, J. H., and T. G. Whitham. 1991. Competition mediating the outcome of a mutualism – protective services of ants as a limiting resource for membracids. *American Naturalist* **138**: 851–865.

Cushman, J. H., J. H. Lawton, and B. F. J. Manly. 1993. Latitudinal patterns in European ant assemblages variation is species richness and body size. *Oecologia* **95**:30–37.

Cushman, J. H., V. K. Rashbrook, and A. J. Beattie. 1994. Assessing benefits to both participants in a lycaenid-ant association. *Ecology* **75**:1031–1041.

Darwin, C. 1890. *The Origin of Species*, 6th edition. London: John Murray.

Davidson, D. W. 1997. The role of resource imbalances in the evolutionary ecology of tropical arboreal ants. *Biological Journal of the Linnean Society* **61**:153–181.

Davidson, D. W. 1998. Resource discovery versus resource domination in ants: a functional mechanism for breaking the trade-off. *Ecological Entomology* **23**:484–490.

Davidson, D. W. 2005. Ecological stoichiometry of ants in a New World rain forest. *Oecologia* **142**:221–231.

Davidson, D. W., S. C. Cook, and R. R. Snelling. 2004. Liquid-feeding performances of ants (Formicidae): ecological and evolutionary implications. *Oecologia* **139**:255–266.

Davidson, D. W., S. C. Cook, R. R. Snelling, and T. H. Chua. 2003. Explaining the abundance of ants in lowland tropical rainforest canopies. *Science* **300**:969–972.

Dean, A. M. 1983. A simple model of mutualism. *American Naturalist* **121**:409–417.

DeAngelis, D. L., and J. C. Waterhouse. 1987. Equilibrium and nonequilibrium concepts in ecological models. *Ecological Monographs* **57**: 1–21.

Delabie, J. H. C. 2001. Trophobiosis between Formicidae and Hemiptera (Sternorrhyncha and Auchenorrhyncha): an overview. *Neotropical Entomology* **30**:501–516.

Del-Claro, K., and P. S. Oliveira. 1993. Ant-homoptera interaction: do alternative sugar sources distract tending ants? *Oikos* **68**: 202–206.

Del-Claro, K., and P. S. Oliveira. 2000. Conditional outcomes in a neotropical treehopper-ant association: temporal and species-specific variation in ant protection and homopteran fecundity. *Oecologia* **124**:156–165.

Denno, R. F. 1994. The evolution of dispersal polymorphism in insects: the influence of habitats, host plants and mates. *Research in Population Ecology Kyoto* **36**:127–135.

Denno, R. F., C. Gratton, H. Dobel, and D. L. Finke. 2003. Predation risk affects relative strength of top-down and bottom-up impacts on insect herbivores. *Ecology* **84**:1032–1044.

Denno, R. F., C. Gratton, M. A. Peterson, G. A. Langellotto, D. L. Finke, and A. F. Huberty. 2002. Bottom-up forces mediate natural-enemy impact in a phytophagous insect community. *Ecology* **83**:1443–1458.

Denno, R. F., K. L. Olmstead, and E. S. McCloud. 1989. Reproductive cost of flight capability: a comparison of life-history traits in wing dimorphic planthoppers. *Ecological Entomology* **14**:31–44.

Denno, R. F., M. A. Peterson, C. Gratton, J. A. Cheng, G. A. Langellotto, A. F. Huberty, and D. L. Finke. 2000. Feeding-induced changes in plant quality mediate interspecific competition between sap-feeding herbivores. *Ecology* **81**:1814–1827.

Denno, R. F., G. K. Roderick, K. L. Olmstead, and H. G. Dobel. 1991. Density-related migration in planthoppers (Homoptera, Delphacidae): the role of habitat persistence. *American Naturalist* **138**:1513–1541.

DeVries, P. J. 1991a. Evolutionary and ecological patterns in myrmecophilous riodinid butterflies. In *Ant-plant Interactions*, ed. C. R. Huxley and D. F. Cutler, pp. 143–156. Oxford: Oxford University Press.

DeVries, P. J. 1991b. Mutualism between *Thisbe irenea* butterflies and ants, and the role of ant ecology in the evolution of larval ant associations. *Biological Journal of the Linnean Society* **43**:179–195.

DeVries, P. J. 1997. *The Butterflies of Costa Rica and their Natural History*. vol. II, *The Riodinidae*. Princeton: Princeton University Press.

Diamond, J. M. 1978. Niche shifts and the rediscovery of interspecific competition. *American Scientist* **66**:322–331

Dietrich, C. H., and L. L. Deitz. 1993. Superfamily Membracoidea (Homoptera, Auchenorrhyncha). 2. Cladistic-Analysis and Conclusions. *Systematic Entomology* **18**:297–311.

Dixon, A. F. G. 1958. The escape response shown by certain aphids to the presence of the coccinellid *Adalia decempunctata* (L.). *Transactions of the Royal Entomological Society of London* **110**:319–334.

Dixon, A. F. G. 1975. Effect of population density and food quality on autumnal reproductive activity in the sycamore aphid, *Drepanosiphum platanoides* (Schr.). *Journal of Animal Ecology* **44**:297–304.

Dixon, A. F. G. 1984. Plant architectural complexity and alary polymorphism in tree-dwelling aphids. *Ecological Entomology* **9**:117–118.

Dixon, A. F. G. 1998. *Aphid Ecology*, 2nd edition. London: Chapman and Hall.

Dixon, A. F. G. 2000. *Insect Predator-Prey Dynamics: Ladybird Beetles and Biological Control*. Cambridge: Cambridge University Press.

Dixon, A. F. G. 2005. *Insect Herbivore-Host Dynamics: Tree-Dwelling Aphids*. Cambridge: Cambridge University Press.

Dixon, A. F. G., and B. K. Agarwala. 1999. Ladybird-induced life-history changes in aphids. *Proceedings of the Royal Society of London Series B* **266**:1549–1553.

Dixon, A. F. G., and P. Kindlmann. 1998. Generation time ratio and the effectiveness of ladybirds as classical biological control agents. In *Pest Management – Future Challenges*, ed. M. P. Zalucki, R. A. I. Drew, and G. G. White pp. 314–320. Proceedings of the 6th Australasian Applied Entomology and Research Conference. Brisbane: University of Queensland.

Dixon, A. F. G., and S. McKay. 1970. Aggregation in the sycamore aphid *Drepanosiphum platanoides* (Schr.) (Hemiptera: Aphididae) and its relevance to the regulation of population growth. *Journal of Animal Ecology* **39**:439–454.

Dixon, A. F. G., M. D. Burns, and S. Wangboonkong. 1968. Migration in aphids: response to current adversity. *Nature* **220**:1337–1338.

Dixon, A. F. G., S. Horth, and P. Kindlmann. 1993. Migration in insects: cost and strategies. *Journal of Animal Ecology* **62**:182–190.

Dixon, A. F. G., P. Kindlmann, J. Leps, and J. Holman. 1987. Why there are so few species of aphids, especially in the tropics? *American Naturalist* **129**:580–592.

Doebeli, M., and N. Knowlton. 1998. The evolution of interspecific mutualism. *Proceedings of the National Academy of Sciences of the USA* **95**:8676–8680.

Donovan, T., and C. W. Welden. 2001. *Spreadsheet Exercises in Ecology and Evolution*. Sunderland, MA: Sinauer Associates.

Douglas, A. E. 1989. Mycetocyte symbiosis in insects. *Biological Review* **64**:409–434.

Douglas, A. E. 1998. Nutritional interactions in insect-microbial symbioses: aphids and their symbiotic bacteria *Buchnera*. *Annual Review of Entomology* **43**:17–37.

Douglas, A. E. 2003. The nutritional physiology of aphids. *Advances in Insect Physiology* **31**:73–140.

Douglas, A. E., and W. A. Prosser. 1992. Synthesis of the essential amino acid tryptophan in the pea aphid (*Acyrthosiphon pisum*) symbiosis. *Journal of Insect Physiology* **38**:565–568.

Douglas, A. E., L. B. Minto, and T. L. Wilkinson. 2001. Quantifying nutrient production by the microbial symbionts in an aphid. *Journal of Experimental Biology* **204**:349–358.

Dugatkin, L. A. 2002. Cooperation in animals: an evolutionary overview. *Biology and Philosophy* **17**:459–476.

Dugatkin, L. A. and M. Mesterton-Gibbons. 1996. Cooperation among unrelated individuals: reciprocal altruism, by-product mutualism and group selection in fishes. *BioSystems* **37**:19–30.

Dugatkin, L. A., and H. K. Reeve. 1994. Behavioral ecology and level of selection: dissolving the group selection controversy. *Advances in the Study of Behaviour* **23**:101–133.

Eastop, V. F. 1973. Deductions from the present day host plants of aphids and related insects. In *Insect-plant Relationships*, ed. H. F. van Emden, pp. 157–178. 6th Symposium, Royal Entomological Society of London. Oxford: Blackwell.

Eastwood, R., and A. M. Fraser. 1999. Associations between lycaenid butterflies and ants in Australia. *Australian Ecology* 24:503–537.

Edson, J. 1985. The influence of predation and resource subdivision on the coexistence of goldenrod aphids. *Ecology* 66:1736–1743.

Eisner, T. 1957. A comparative morphological study of the proventriculus of ants (Hymenoptera: Formicidae). *Bulletin of the Museum of Comparative Zoology* 116: 429–490.

Eliot, J. N. 1973. The higher classification of the Lycaenidae (Lepidoptera): a tentative arrangement. *Bulletin of the British Museum of Natural History* 28:371–505.

Elmes, G. W., J. A. Thomas, M. L. Munguira, and K. Fiedler. 2001. Larvae of lycaenid butterflies that parasitize ant colonies provide exceptions to normal insect growth rules. *Biological Journal of the Linnean Society* 73:259–278.

El-Ziady, S., and J. S. Kennedy. 1956. Beneficial effects of the common garden ant, *Lasius niger* L. on the black bean aphid, *Aphis fabae* Scopoli. *Proceedings of the Royal Entomological Society of London A.* 31:61–65.

Engel, V., M. K. Fischer, F. L. Wäckers, and W. Völkl. 2001. Interactions between extrafloral nectaries, aphids and ants: are there competition effects between plant and homopteran sugar sources? *Oecologia* 129: 577–584.

Ewald, P. W. 1994. *Evolution of Infectious Disease*. Oxford: Oxford University Press.

Ewart, W. H., and R. L. Metcalf. 1956. Preliminary studies of sugar and amino acids in the honeydew of five species of coccids feeding on citrus in California. *Annals of the Entomological Society of America* 49:441–447.

Feener, D. H., Jr. 1981. Competition between ant species: outcome controlled by parasitic flies. *Science* 214:815–817.

Feinsinger, P., E. E. Spears, and R. W. Poole. 1981. A simple measure of niche breadth. *Ecology* 62:27–32.

Fellers, J. H. 1987. Interference and exploitation in a guild of woodland ants. *Ecology* 68:1466–1478.

Ferriere, R., J. L. Bronstein, S. Rinaldi, R. Law, and M. Gauduchon. 2002. Cheating and the evolutionary stability of mutualisms. *Proceedings of the Royal Entomological Society of London* 269:773–780.

Fiedler, K. 1991. Systematic, evolutionary, and ecological implications of myrmecophily within Lycaenidae (Insecta: Lepidoptera: Papilionoidea). *Bonner Zoological Monograph* 31:5–157.

Fiedler, K. 1994. Lycaenid butterflies and plants: is myrmecophily associated with amplified host plant diversity? *Ecological Entomology* 19: 79–82.

Fiedler, K. 1995. Lycaenid butterflies and plants: is myrmecophily associated with particular host plant preferences? *Ethology Ecology and Evolution* 7: 107–132.

Fiedler, K. 1996. Host-plant relationships of lycaenid butterflies: large-scale patterns, interactions with plant chemistry, and mutualism with ants. *Entomologia Experimentalis et Applicata* 80:259–267.

Fiedler, K. 1997a. Geographical patterns in life-history traits of Lycaenidae butterflies – ecological and evolutionary implications. *Zoology* 100:336–347.

Fiedler, K. 1997b. Life-history patterns of myrmecophilous butterflies and other insects: their implications on tropical species diversity. In *Tropical Biodiversity and Systematics*, ed. H. Ulrich, pp. 71–92. Bonn: Zoologisches Forschungsinstitut und Museum Alexander Koenig.

Fiedler, K. 2001. Ants that associate with Lycaenidae butterfly larvae: diversity, ecology and biogeography. *Diversity and Distributions* **7**:45–60.

Fiedler, K., B. Holldobler, and P. Seufert. 1996. Butterflies and ants: the communicative domain. *Experientia* **52**:14–24.

Fischer, M. K., and A. W. Shingleton. 2001. Host plant and ants influence the honeydew sugar composition of aphids. *Functional Ecology* **15**:544–550.

Fischer, M. K., K. H. Hoffmann, and W. Völkl. 2001. Competition for mutualists in an ant-homopteran interaction mediated by hierarchies of ant-attendance. *Oikos* **92**:531–541.

Fischer, R. C., S. M. Olzant, W. Wanek, and V. Mayer. 2005. The fate of *Corydalis cava* elaiosomes within an ant colony of *Myrmica rubra*: elaiosomes are preferentially fed to larvae. *Insectes Sociaux* **52**:55–62.

Fisher, D. B., J. P. Wright, and T. E. Mittler. 1984. Osmoregulation by the aphid *Myzus persicae*: a physiological role for honeydew oligosaccharides. *Journal of Insect Physiology* **30**:387–393.

Flanders, S. E. 1957. The complete interdependence of an ant and a coccid. *Ecology* **38**:535–536.

Flatt, T., and W. W. Weisser. 2000. The effects of mutualistic ants on aphid life history traits. *Ecology* **81**:3522–3529.

Fowler, H. G. 1993. Differential recruitment in *Camponotus rufipes* (Hymenoptera: Formicidae) to protein and carbohydrate resources. *Naturalia* **18**:9–13.

Fraser, A. M., A. H. Axen, and N. E. Pierce. 2001. Assessing the quality of different ant species as partners of a myrmecophilous butterfly. *Oecologia* **129**:452–460.

Fraser, A. M., T. Tregenza, N. Wedell, M. A. Elgar, and N. E. Pierce. 2002. Oviposition tests of ant preference in a myrmecophilous butterfly. *Journal of Evolutionary Biology* **15**:861–870.

Freitas, A. V. L., and P. S. Oliveira. 1992. Biology and behaviour of the neotropical butterfly *Eunica bechina* (Nymphalidae) with special reference to larval defence against ant predation. *Journal of Research on the Lepidoptera* **31**:1–11.

Freitas, A. V. L., and P. S. Oliveira. 1996. Ants as selective agents on herbivore biology: effects on the behaviour of a non-myrmecophilous butterfly. *Journal of Animal Ecology* **65**:205–210.

Gaume, L., D. Matile-Ferrero, and D. McKey. 2000. Colony formation and acquisition of coccoid trophobionts by *Aphomomyrmex afer* (Formicinae): co-dispersal of queens and phoretic mealybugs in an ant-plant-homopteran mutualism? *Insectes Sociaux* **47**: 84–91.

Gaume, L., D. McKey, and S. Terrin. 1998. Ant-plant-homopteran mutualism: how the third partner affects the interaction between a plant-specialist ant and its myrmecophyte host. *Proceedings of the Royal Society of London Series B* **265**:569–575.

Gish, M., and M. Inbar. 2006. Host location by apterous aphids after escape dropping from the plant. *Journal of Insect Behavior* **19**: 143–153.

Goidanich, A. 1956. *Stomaphis quercus* and ants. *Bulletin of the Institite of Entomology of the University of Bologna* **23**:93–131.

Gonzales, W. L., E. Fuentes-Contreras, and H. M. Niemeyer. 2002. Host plant and natural enemy impact on cereal aphid competition in a seasonal environment. *Oikos* **96**:481–491.

Goodchild, A. J. P. 1966. Evolution of the alimentary canal in the Hemiptera. *Biological Reviews* **41**:97–140.

Gösswald, K. 1938. Über den Einfluß von verschiedenen Temperaturen und Luftfeuchtigkeit auf die Lebensäusserungen der Ameisen. 1. Die Lebensdauer ökologisch verschiedener Ameisenarten unter dem Einfluß bestimmter Luftfeuchtigkeit und Temperatur. *Zeitschrift für wissenschaftliche Zoologie* (**A**):247–344.

Gösswald, K. 1941. Rassenstudien an der roten Waldameise *Formica rufa* L. auf systematischer, ökologischer, physiologischer und biologischer Grundlage. *Zeitschrift für angewandte Entomologie* **28**:62–124.

Gösswald, K. 1989a. *Die Waldameise im Ökosystem Wald, Nutzen und Hege.* Wiesbaden: Aula Verlag.

Gösswald, K. 1989b. *Die Waldameise: Biologische Grundlagen, Ökologie und Verhalten.* Wiesbaden: Aula Verlag.

Gotelli, N. J. 1996. Ant community structure: effects of predatory ant lions. *Ecology* **77**:630–638.

Gotelli, N. J., and A. E. Arnett. 2000. Biogeographic effects of red fire ant invasion. *Ecology Letters* **3**:257–261.

Gotelli, N. J., and A. M. Ellison. 2002. Assembly rules for New England ant assemblages. *Oikos* **99**:591–599.

Gould, S. J. 1988 Kropotkin was no crackpot. *Natural History* **7**: 12–21.

Graham, M. H., and P. K. Dayton. 2002. On the evolution of ecological ideas: paradigms and scientific progress. *Ecology* **83**:1481–1489.

Grassé, P.-P. 1951. *Traité de Zoologie: Anatomie, Systematique, Biologie.* Tome X, Fascicule II, *Insectes Supérieurs et Hémiptéroïdes.* Paris: Masson.

Greathead, D. J. 1990. Crawler behaviour and dispersal. In *World Crop Pests. Armored Scale Insects: Their Biology, Natural Enemies and Control*, ed. D. Rosen, pp. 305–308. Amsterdam: Elsevier.

Greene, C. M. 2003. Habitat selection reduces extinction of populations subject to Allee effects. *Theoretical Population Biology* **64**:1–10.

Greene, C. M., and J. A. Stamps. 2001. Habitat selection at low population densities. *Ecology* **82**:2091–2100.

Gruppe, A., and P. Römer. 1988. The lupin aphid (*Macrosiphum albifrons* Essig, 1911) (Hom, Aphididae) in West Germany: its occurrence, host plants and natural enemies. *Journal of Applied Entomology* **106**:135–143.

Gullan, P. 1997. Relationships with ants. In *Soft Scale Insects: Their Biology, Natural Enemies and Control*, ed. Y. Ben-Dov and C. J. Hodgson, pp. 351–373. Amsterdam: Elsevier.

Gullan, P. J., and M. Kosztarab. 1997. Adaptations in scale insects. *Annual Review of Entomology* **42**:23–50.

Gullan, P., and J. H. Martin. 2003. Sternorrhyncha (jumping plant lice, whiteflies, aphids, and scale insects). In *Encyclopedia of Insects*, ed. V. H. Resh and R. T. Crade, pp. 1079–1089. Amsterdam: Elsevier Academic Press.

Hairston, G., F. E. Smith, and L. B. Slobodkin. 1960. Community structure, population control, and competition. *American Naturalist* **64**:421–425.

Hajek, A. E., and D. L. Dahlsten. 1986. Coexistence of three species of leaf-feeding aphids (Homoptera) on *Betula pendula.* *Oecologia* **68**:380–386.

Hale, B. K., J. S. Bale, J. Pritchard, G. J. Masters, and V. K. Brown. 2003. Effects of host plant drought stress on the performance of the bird cherry-oat aphid, *Rhopalosiphum padi* (L.): a mechanistic analysis. *Ecological Entomology* **28**:666–677.

Hamilton, W. D., and R. M. May. 1977. Dispersal in stable habitats. *Nature* **269**:578–581.

Hanks, L. M., and R. F. Denno. 1993. The role of demic adaptation in colonization and spread of scale insect populations. In *Evolution of Insect Pests: Patterns of Variation*, ed. K. C. Kim and B. A. McPheron, pp. 393–411. New York: Wiley & Sons.

Hanski, I. 1998. *Metapopulation Ecology*. Oxford: Oxford University Press.

Hanski, I., and M. E. Gilpin. 1997. *Metapopulation Biology: Ecology, Genetics and Evolution*. London: Academic Press.

Harmon, J. P., and D. A. Andow. 2007. Behavioral mechanism underlying ants' density-dependent deterence of aphid-eating predators. *Oikos* **116**: 1030–1036.

Hanski, I., and I. P. Woiwod. 1993. Spatial synchrony in the dynamics of moth and aphid populations. *Journal of Animal Ecology* **62**:656–668.

Hay, M. E., J. D. Parker, D. E. Burkepile, C. C. Caudill, A. E. Wilson, Z. P. Hallinan, and A. D. Chequer. 2004. Mutualism and aquatic community structure: the enemy of my enemy is my friend. *Annual Review of Ecology, Evolution, and Systematics* **35**:175–197.

Hayamizu, E. 1982. Comparative studies on aggregations among aphids in relation to population dynamics. 1. Colony formation and aggregation behavior of *Brevicoryne brassicae* L. and *Myzus persicae* (Sulzer) (Homoptera, Aphididae). *Applied Entomology and Zoology* **17**:519–529.

Heinsohn, R., and C. Packer. 1995. Complex cooperative strategies in group-territorial African lions. *Science* **269**:1260–1262.

Heinze, J. 1995. Reproductive skew and genetic relatedness in *Leptothorax* ants. *Proceedings of the Royal Society of London Series B* **261**:375–379.

Heithaus, E. R., D. C. Culver, and A. J. Beattie. 1980. Models of some ant-plant mutualisms. *American Naturalist* **16**:347–361.

Helms, K. R., and S. B. Vinson. 2002. Widespread association of the invasive ant *Solenopsis invicta* with an invasive mealybug. *Ecology* **83**:2425–2438.

Hennig, W. 1969. Die Stammesgeschichte der Insekten. Frankfurt a.Main: Senkenberg, Naturforschende Gesellschaft.

Herre, E. A., N. Knowlton, U. G. Mueller, and S. A. Rehner. 1999. The evolution of mutualisms: exploring the paths between conflict and cooperation. *Trends in Ecology and Evolution* **14**:49–53.

Hill, M. G., and P. J. M. Blackmore. 1980. Interactions between ants and the coccid *Icerya seychellarum* on Aldabra Atoll. *Oecologia* **45**:360–365.

Hixon, M. A., S. W. Pacala, and S. A. Sandin. 2002. Population regulation: historical context and contemporary challenges of open vs. closed systems. *Ecology* **83**:1490–1508.

Hochberg, M. E., R. T. Clarke, G. W. Elmes, and J. A. Thomas. 1994. Population dynamic consequences of direct and indirect interactions involving a large blue butterfly and its plant and red ant hosts. *Journal of Animal Ecology* **63**:375–391.

Hochberg, M. E., R. Gomulkiewicz, R. D. Holt, and J. N. Thompson. 2000. Weak sinks could cradle mutualistic symbioses – strong sources should harbour parasitic symbioses. *Journal of Evolutionary Biology* **13**:213–222.

Hoeksema, J. D., and E. M. Bruna. 2000. Pursuing the big questions about interspecific mutualism: a review of theoretical approaches. *Oecologia* **125**:321–330.

Hoeksema, J. D., and M. W. Schwartz. 2002. Expanding comparative-advantage biological market models: contingency of mutualism on partners' resource requirements and acquisition trade-offs. *Proceedings of the Royal Entomological Society of London Series B.* **270**:913–990.

Hol, W. H. G., and J. A. Van Veen. 2002. Pyrrolizidine alkaloids from *Senecio jacobaea* affect fungal growth. *Journal of Chemical Ecology* **28**:1763–1772.

Hol, W. H. G., K. Vrieling, and J. A. van Veen. 2003. Nutrients decrease pyrrolizidine alkaloid concentrations in *Senecio jacobaea*. *New Phytologist* **158**:175–181.

Holland, J. N., D. L. DeAngelis, and J. L. Bronstein. 2002. Population dynamics and mutualism: functional responses of benefits and costs. *American Naturalist* **159**:231–244.

Holland, J. N., J. H. Ness, A. Boyle, and J. L. Bronstein. 2005. Mutualisms as consumer-resource interactions. In *Ecology of Predator-Prey Interactions* ed. P. Barbosa and I. Castellanos. Oxford: Oxford University Press. pp. 17–33.

Hölldobler, B., and E. O. Wilson. 1990. *The Ants*. Cambridge, MA: Harvard University Press.

Holt, R. D. 1977. Predation, apparent competition, and structure of prey communities. *Theoretical Population Biology* **12**:197–229.

Holt, R. D. 2002. Food webs in space: on the interplay of dynamic instability and spatial processes. *Ecological Research* **17**:261–273.

Holt, R. D., and J. H. Lawton. 1994. The ecological consequences of shared natural enemies. *Annual Review of Ecology and Systematics* **25**:495–520.

Holway, D. A., and A. V. Suarez. 2004. Colony-structure variation and interspecific competitive ability in the invasive Argentine ant. *Oecologia* **138**:216–222.

Holway, D. A., L. Lach, A. V. Suarez, N. D. Tsutsui, and T. J. Case. 2002a. The causes and consequences of ant invasions. *Annual Review of Ecology and Systematics* **33**:181–233.

Holway, D. A., A. V. Suarez, and T. J. Case. 2002b. Role of abiotic factors in governing susceptibility to invasion: a test with argentine ants. *Ecology* **83**:1610–1619.

Honek, A. 1991. Environment stress, plant quality and abundance of cereal aphids (Hom., Aphididae) on winter wheat. *Journal of Applied Entomology* **112**:65–70.

Hopkins, G. W., and A. F. G. Dixon. 1997. Enemy-free space and the feeding niche of an aphid. *Ecological Entomology* **22**:271–274.

Hopkins, G. W., and J. I. Thacker. 1999. Ants and habitat specificity in aphids. *Journal of Insect Conservation* **3**:25–31.

Hopkins, G. W., J. I. Thacker, and A. F. G. Dixon. 1998. Limits to the abundance of rare species: an experimental test with a tree aphid. *Ecological Entomology* **23**:386–390.

Howe, H. F. 1984. Constraints on the evolution of mutualism. *American Naturalist* **123**:764–777.

Hsiao, T. C. 1973. Plant responses to water stress. *Annual Review of Plant Physiology* **24**:519–570.

Hubbell, S. P. 2001. *The Unified Neutral Theory of Biodiversity and Biogeography*. Princeton: Princeton University Press.

Huberty, A. F., and R. F. Denno. 2004. Plant water stress and its consequences for herbivorous insects: a new synthesis. *Ecology* **85**:1383–1398.

Hunter, M. D. 2002. Maternal effects and the population dynamics of insects on plants. *Agricultural and Forest Entomology* **4**:1–9.

Hunter, M. D., and P. W. Price. 1998. Cycles in insect populations: delayed density dependence or exogenous driving variables? *Ecological Entomology* **23**:216–222.

Hunter, M. D., and P. W. Price. 2000. Detecting cycles and delayed density dependence: a reply to Turchin and Berryman. *Ecological Entomology* **25**:122–124.

Huxley, T. H. 1888. The struggle for existence in human society. *Nineteenth Century*.

Inbar, M., H. Doostdar, and R. T. Mayer. 2001. Suitability of stressed and vigorous plants to various insect herbivores. *Oikos* **94**:228–235.

Ingram, K. K. 2002a. Flexibility in nest density and social structure in invasive populations of the Argentine ant, *Linepithema humile*. *Oecologia* **133**:492–500.

Ingram, K. K. 2002b. Plasticity in queen number and social structure in the invasive Argentine ant (*Linepithema humile*). *Evolution* **56**:2008–2016.

Ives, A. R., P. Kareiva, and R. Perry. 1993. Responses of a predator to variation in prey density at three hierarchical scales: lady beetles feeding on aphids. *Ecology* **74**:1929–1938.

Janzen, D. H. 1977. What are dandelions and aphids? *American Naturalist* **111**: 586–589.

Janzen, D. H. 1985. The natural history of mutualisms. In *The Biology of Mutualism: Ecology and Evolution*, ed. D. H. Boucher, pp. 40–99. New York: Oxford University Press.

Johnson, B. 1959. Ants and form reversal in aphids. *Nature* **184**:740–741.

Johnson, S. N., A. E. Douglas, S. Woodward, and S. E. Hartley. 2003a. Microbial impacts on plant-herbivore interactions: the indirect effects of a birch pathogen on a birch aphid. *Oecologia* **134**:388–396.

Johnson, S. N., D. A. Elston, and S. E. Hartley. 2003b. Influence of host plant heterogeneity on the distribution of a birch aphid. *Ecological Entomology* **28**:533–541.

Johnstone, R. A., and R. Bshary. 2002. From parasitism to mutualism: partner control in asymmetric interactions. *Ecology Letters* **5**:634–639.

Jordano, D., J. Rodriguez, C. D. Thomas, and J. F. Haeger. 1992. The distribution and density of a lycaenid butterfly in relation to *Lasius* ants. *Oecologia* **91**:439–446.

Jordano, D., and C. D. Thomas. 1992. Specificity of an ant-lycaenid interaction. *Oecologia* **91**:431–438.

Kainulainen, P., J. Holopainen, V. Palomäki, and T. Holopainen. 1996. Effects of nitrogen fertilization on secondary chemistry and ectomycorrhizal state of Scots pine seedlings and on growth of grey pine aphid. *Journal of Chemical Ecology* **22**:617–636.

Kaneko, S. 2002. Aphid-attending ants increase the number of emerging adults of the aphid's primary parasitoid and hyperparasitoids by repelling intraguild predators. *Entomological Science* **5**:131–146.

Kaplan, I., and M. D. Eubanks. 2002. Disruption of cotton aphid (Homoptera: Aphididae) – Natural enemy dynamics by red imported fire ants (Hymenoptera: Formicidae). *Environmental Entomology* **31**:1175–1183.

Kareiva, P. 1987. Habitat fragmentation and the stability of predator-prey interactions. *Nature* **326**:388–390.

Karsai, I., and J. W. Wenzel. 1998. Productivity, individual-level and colony-level flexibility, and organization of work as consequences of colony size. *Proceedings of the National Academy of Sciences* **95**:8665–8669.

Kaspari, M., and E. L. Vargo. 1995. Colony size as a buffer against seasonality – Bergmanns rule in social insects. *American Naturalist* **145**:610–632.

Kaspari, M., L. Alonso, and S. O'Donnell. 2000a. Three energy variables predict ant abundance at a geographical scale. *Proceedings of the Royal Society of London Series B* **267**:485–489.

Kaspari, M., S. O'Donnell, and J. R. Kercher. 2000b. Energy, density, and constraints to species richness: ant assemblages along a productivity gradient. *American Naturalist* **155**:280–293.

Katayama, N., and N. Suzuki. 2002. Cost and benefit of ant-attendance for *Aphis craccivora* (Hemiptera: Aphididae) with reference to aphid colony size. *Canadian Entomologist* **134**:241–249.

Kay, A. 2004. The relative availabilities of complementary resources affect the feeding preferences of ant colonies. *Behavioral Ecology* **15**:63–70.

Keddy, P. 1990. Is mutualism really irrelevant to ecology? *Bulletin of the Ecological Society of America* **71**:101–102.

Keeler, K. H. 1979. Distribution of ants with extrafloral nectaries and ants at two elevations in Jamaica. *Biotropica* **11**:152–154.

Keeler, K. H. 1981. A model of selection for facultative nonsymbiotic mutualism. *American Naturalist* **118**:488–498.

Keeler, K. H. 1985. Extrafloral nectaries on plants in communities without ants: Hawaii. *Oikos* **44**:407–414.

Keller, L., and M. Chapuisat. 1999. Cooperation among selfish individuals in insect societies. *BioScience* **49**:899–909.

Keller, L., and H. K. Reeve. 1994. Partitioning of reproduction in animal societies. *Trends in Ecology and Evolution* **9**:98–102.

Kennedy, J. S. and C. O. Booth 1959. Responses of *Aphis fabae* Scop to water shortage in host plants in the field. *Entomologia Experimentalis et Applicata* **2**:1–11.

Kennedy, J. S., K. P. Lamb and C. O. Booth. 1958. Responses of *Aphis fabae* Scop. to water shortage in host plants in pots. *Entomologia Experimentalis et Applicata* **1**: 274–279.

Killingback, T., M. Doebeli, and N. Knowlton. 1999. Variable investment, the Continuous Prisoner's Dilemma, and the origin of cooperation. *Proceedings of the Royal Society of London Series B* **266**:1723–1728.

Kindlmann, P. and A. F. G. Dixon. 1999. Generation time ratios – determinants of prey abundance in insect predator–prey interactions. *Biological Control* **16**:133–138.

Kindlmann, P., M. Hulle, and B. Stadler. 2007. Timing of dispersal: effects of ants on aphids. *Oecologia* **152**:625–631.

Kingsland, S. 1995. *Modeling Nature: Episodes in the History of Population Ecology*, 2nd edition. Chicago: University of Chicago Press.

Kiss, A. 1981. Melezitose, aphids and ants. *Oikos* **37**:382.

Kitching, R. L. 1981. Egg clustering and the southern hemisphere lycaenids: comments. *American Naturalist* **118**:423–425.

Kloft, W. J. 1959. Versuch einer Analyse der trophobiotischen Beziehungen von Ameisen zu Aphiden. *Biologisches Zentralblatt* **78**:863–870.

Kneitel, J. M., and J. M. Chase. 2004. Trade-offs in community ecology: linking spatial scales and species coexistence. *Ecology Letters* **7**:69–80.

Koptur, S. 1991. Extrafloral nectaries of herbs and trees: modeling the interaction with ants and parasitoids. In *Ant-plant Interactions*, ed. C. R. Huxley and D. F. Cutler, pp. 213–230. Oxford: Oxford University Press.

Koricheva, J., S. Larsson, and E. Haukioja. 1998. Insect performance on experimentally stressed woody plants: a meta-analysis. *Annual Review of Entomology* **43**:195–216.

Koteja, J. 1985. Essay on the prehistory of the scale insects (Homoptera, Coccinea). *Annales Zoologici (Wars.)* **38**:461–504.

Kropotkin, P. A. 1902. *Mutual Aid: A Factor of Evolution*. London: William Heinemann. Also, 1998. London: Freedom Press.

Kruess, A., and T. Tscharntke. 1994. Habitat fragmentation, species loss, and biological control. *Science* **264**:1581–1584.

Kundu, R., and A. F. G. Dixon. 1995. Evolution of complex life cycles in aphids. *Journal of Animal Ecology* **64**:245.

Kunert, G., and W. W. Weisser. 2003. The interplay between density- and trait mediated effects in predator-prey interactions: a case study in aphid wing polymorphism. *Oecologia* **135**:304–312.

Kunkel, H., W. J. Kloft, and A. Fossel. 1985. Die Honigtau-Erzeuger des Waldes. In *Waldtracht und Waldhonig in der Imkerei*, ed. W. J. Kloft and H. Kunkel, pp. 48–265. Munich: Ehrenwirth.

Labandeira, C. C. 1997. Insect mouthparts: ascertaining the paleobiology of insect feeding strategies. *Annual Review of Ecology and Systematics* **28**:153–193.

Larsson, S. 1989. Stressful times for the plant-insect performance hypothesis. *Oikos* **56**:277–283.

Larsson, S., and C. Bjorkman. 1993. Performance of chewing and phloem-feeding insects on stressed trees. *Scandinavian Journal of Forest Research* **8**:550–559.

Lees, A. D. 1967. The production of the apterous and alatae forms in the aphid *Megoura viciae* (Buckton), with special reference to the role of crowding. *Journal of Insect Physiology* **13**:289–318.

Leibold, M. A., M. Holyoak, N. Mouquet *et al.* 2004. The metacommunity concept: a framework for multi-scale community ecology. *Ecology Letters* **7**:601–613.

Leimar, O., and A. H. Axen. 1993. Strategic behavior in an interspecific mutualism: interactions between lycaenid larvae and ants. *Animal Behaviour* **46**:1177–1182.

Levieux, J. 1977. La nutrition des fourmis tropicales – V. Eléments de synthèse. Les modes d'exploitation de la biocoenose. *Insectes Sociaux* **24**:235–260.

Levieux, J., and D. Louis. 1975. La nutrition des fourmis tropicales – II. Comportement alimentaire et régime de *Camponotus vividus* (Smith) (Hymenoptera Formicidae). *Insectes Sociaux* **22**:391–404.

Levin, S. A. 1992. The problem of pattern and scale in ecology. *Ecology* **73**:1943–1967.

Levins, R., and D. Culver. 1971. Regional coexistence of species and competition between rare species. *Proceedings of the National Academy of Sciences* **68**:1246–1248.

Lin, C. P., B. N. Danforth, and T. K. Wood. 2004. Molecular phylogenetics and evolution of maternal care in Membracine treehoppers. *Systematic Biology* **53**:400–421.

Loreau, M. 1995. Consumers as maximizers of matter and energy flow in ecosystems. *American Naturalist* **145**:22–42.

Loreau, M., N. Mouquet, and R. D. Holt. 2003. Meta-ecosystems: a theoretical framework for a spatial ecosystem ecology. *Ecology Letters* **6**:673–679.

Lotka, A. J. 1925. *Elements of Physiological Biology*. New York: Dover Publication (1956).

Mackauer, M., and W. Völkl. 1993. Regulation of aphid populations by aphid wasps: does parasitoid foraging behaviour or hyperparasitism limit impact? *Oecologia* **94**:339–350.

Malicky, H. 1969. Versuch einer Analyse der ökologischen Beziehungen zwischen Lycaeniden (Lepidoptera) und Formiciden (Hymenoptera). *Tijdschrift voor Entomologie* **112**:85–90.

Malicky, H. 1970. New aspects of the association between lycaenid larvae (Lycaenidae) and ants (Formicidae; Hymenoptera). *Journal of the Lepidopterists' Society* **24**:190–202.

Maschwitz, U., K. Dumpert, and K. R. Tuck. 1986. Ants feeding on anal exudate from tortricid larvae: a new type of trophobiosis. *Journal of Natural History* **20**:1041–1050.

Maschwitz, U., B. Fiala, and W. R. Dolling. 1987. New trophobiotic symbioses of ants with South-East-Asian bugs. *Journal of Natural History* **21**:1097–1107.

Mattson, J. W. 1980. Herbivory in relation to plant nitrogen content. *Annual Review of Ecology and Systematics* **11**:119–161.

Mattson, J. W., and R. A. Haack. 1987. The role of drought stress in provoking outbreaks of phytophagous insects. In *Insect Outbreaks: Ecological and Evolutionary Perspectives*, ed. P. Barbosa and J. Schultz, pp. 365–407. New York: Academic Press.

Maurer, B. A. 1999. *Untangling Ecological Complexity: The Macroscopic Perspective.* Chicago: University of Chicago Press.

May, R. M. 1972. Will a larger complex system be stable? *Nature* **238**:413–417.

May, R. M. 1973. *Stability and Complexity in Model Ecosystems.* Princeton: Princeton University Press.

May, R. M. and J. Seger (1986) Ideas in ecology. *American Scientist* **74**: 256–267.

MacArthur, R. H. 1972. *Geographical Ecology.* New York: Harper & Row.

MacArthur, R. H., and O. E. Wilson. 1967. *The Theory of Island Biogeography.* Princeton: Princeton University Press.

McClure, M. S. 1980. Foliar nitrogen: a basis for host suitability for elongate hemlock scale, *Fiorinia externa* (Homoptera, Diaspididae). *Ecology* **61**:72–79.

McEvoy, P. B. 1979. Advantages and disadvantages to group living in treehoppers (Homoptera: Membracidae). *Miscellaneous Publications of the Entomological Society of America* **11**:1–13.

McGlynn, T. P. 1999. The worldwide transfer of ants: geographical distribution and ecological invasions. *Journal of Biogeography* **26**: 535–548.

McIver, J. D., and C. Loomis. 1993. A size-distance relation in Homoptera-tending thatch ants (*Formica obscuripes, Formica planipilis*). *Insectes Sociaux* **40**:207–218.

McKamey, S. H. 1998. Taxonomic catalogue of the Membracoidea (exclusive of leafhoppers): second supplement to Fascicle I: Membracidae of the general catalogue of the Hemiptera. *Memoirs of the American Entomological Institute* **60**:1–377.

McKamey, S. H., and L. L. Deitz. 1996. Generic revision of the new world tribe Hoplophorionini (Hemiptera: Membracidae: Membracinae). *Systematic Entomology* **21**:295–342.

Messina, F. J. 1981. Plant protection as a consequence of an ant-membracid mutualism: interactions on Goldenrod (*Solidago* sp.). *Ecology* **62**: 1433–1440.

Mesterton-Gibbons, M., and L. A. Dugatkin. 1992. Cooperation among unrelated individuals: evolutionary factors. *Quarterly Review of Biology* **67**:267–281.

Miles, P. W., D. Aspinall, and L. Rosenberg. 1982. Performance of the cabbage aphid, *Brevicoryne brassicae* (L), on water-stressed rape plants, in relation to changes in their chemical composition. *Australian Journal of Zoology* **30**:337–345.

Miller, D. R., and M. Kosztarab. 1979. Recent advances in the study of scale insects. *Annual Review of Entomology* **24**:1–27.

Mittler, T. E. 1958. Studies on the feeding and nutrition of *Tuberolachnus salignus* (Gmelin) (Homoptera, Aphididae). 11. The nitrogen and sugar composition of ingested phloem sap and excreted honeydew. *Journal of Experimental Biology.* **35**: 74–84.

Mole, S., and A. J. Zera. 1993. Differential allocation of resources underlies the dispersal-reproduction trade-off in the wing-dimorphic cricket, *Gryllus rubens.* *Oecologia* **93**:121–127.

Mole, S., and A. J. Zera. 1994. Differential resource consumption obviates a potential flight fecundity trade-off in the sand cricket (*Gryllus firmus*). *Functional Ecology* **8**:573–580.

Molyneux, R. J., B. C. Campbell, and D. L. Dreyer. 1990. Honeydew analysis for detecting phloem transport of plant natural products: implications for host plant resistance to sap sucking insects. *Journal of Chemical Ecology* **16**:1899–1909.

Mondor, E. B., B. D. Roitberg, and B. Stadler. 2002. Cornicle length in Macrosiphini aphids: a comparison of ecological traits. *Ecological Entomology* **27**:758–762.

Montllor, C. B. 1991. The influence of plant chemistry on aphid feeding behavior. In *Insect Plant Interactions*, ed. E. Bernays, pp. 125–173. Boston: CRC Press.

Mooney, K. A. and C. V. Tillberg. 2005. Temporal and spatial variation to ant omnivory in pine forests. *Ecology* **86**:1225–1235.

Morales, M. A. 2000a. Mechanisms and density dependence of benefit in an ant-membracid mutualism. *Ecology* **81**:482–489.

Morales, M. A. 2000b. Survivorship of an ant-tended membracid as a function of ant recruitment. *Oikos* **90**:469–476.

Morales, M. A. 2002. Ant-dependent oviposition in the membracid *Publilia concava*. *Ecological Entomology* **27**:247–250.

Moran, N. 1992. The evolution of aphid life cycles. *Annual Review of Entomology* **37**:321–348.

Morris, W. F., J. L. Bronstein, and W. G. Wilson. 2003. Three-way coexistence in obligate mutualist-exploiter interactions: the potential role of competition. *American Naturalist* **161**:860–875.

Morrison, L. W. 2002. Island biogeography and metapopulation dynamics of Bahamian ants. *Journal of Biogeography* **29**:387–394.

Mouquet, N., and M. Loreau. 2002. Coexistence in metacommunities: the regional similarity hypothesis. *American Naturalist* **159**:420–426.

Mouquet, N., and M. Loreau. 2003. Community patterns in source-sink metacommunities. *American Naturalist* **162**:544–557.

Mousseau, T. A., and H. Dingle. 1991. Maternal effects in insect life histories. *Annual Review of Entomology* **36**:511–534.

Mousseau, T. A., and C. W. Fox. 1998. The adaptive significance of maternal effects. *Trends in Ecology and Evolution* **13**:403–407.

Mueller, U. G., T. R. Schultz, C. R. Currie, R. M. M. Adams, and D. Malloch. 2001. The origin of the attine ant-fungus mutualism. *Quarterly Review of Biology* **76**:169–197.

Müller, C. B., and H. C. J. Godfray. 1997. Apparent competition between two aphid species. *Journal of Animal Ecology* **66**:57–64.

Müller, C. B., and H. C. J. Godfray. 1999. Predators and mutualists influence the exclusion of aphid species from natural communities. *Oecologia* **119**:120–125.

Muller-Landau, H. C., S. A. Levin, and J. E. Keymer. 2003. Theoretical perspectives on evolution of long-distance dispersal and the example of specialized pests. *Ecology* **84**:1957–1967.

Murdoch, W. W. 1994. Population regulation in theory and practice. *Ecology* **75**:271–287.

Murdoch, W. W., C. J. Briggs, and R. M. Nisbet. 2003. *Consumer-Resource Dynamics*. Oxford: Princeton University Press.

Murray, B. G. 1999. Can the population regulation controversy be buried and forgotten? *Oikos* **84**:148–152.

Nee, S., and R. M. May. 1992. Dynamics of metapopulations: habitat destruction and competitive coexistence. *Journal of Animal Ecology* **61**:37–40.

Neuhauser, C., and J. E. Fargione. 2004. A mutualism-parasitism continuum model and its application to plant-mycorrhizae interactions. *Ecological Modelling* **177**:337–352.

Nice, C. C., J. A. Fordyce, A. M. Shapiro, and R. Ffrench-Constant. 2002. Lack of evidence for reproductive isolation among ecologically specialised lycaenid butterflies. *Ecological Entomology* **27**:702–712.

Nicholson, A. J. 1933. The balance of animal populations. *Journal of Animal Ecology* **38**:131–178.

Nicholson, A. J., and V. A. Bailey. 1935. The balance of animal populations. *Proceedings of the Zoological Society of London* **3**:551–598.

Nixon, G. E. J. 1951. *The Association of Ants with Aphids and Coccids*. London: Commonwealth Institute of Entomology.

Noe, R., and P. Hammerstein. 1994. Biological markets: supply and demand determine the effect of partner choice in cooperation, mutualism and mating. *Behavioral Ecology and Sociobiology* **35**:1–11.

Noe, R., and P. Hammerstein. 1995. Biological markets. *Trends in Ecology and Evolution* **10**:336–339.

Nonacs, P., and P. Calabi. 1992. Competition and predation risk: their perception alone affects ant colony growth. *Proceedings of the Royal Entomological Society of London Series B* **249**:95–99.

Nonacs, P., and L. M. Dill. 1990. Mortality risk vs. food quality trade-offs in a common currency: ant patch preferences. *Ecology* **71**:1886–1892.

Nonacs, P., and L. M. Dill. 1991. Mortality risk versus food quality trade-offs in ants: patch use over time. *Ecological Entomology* **16**:73–80.

Nowak, M. A., and R. M. May. 1992. Evolutionary games and spatial chaos. *Nature* **359**:826–829.

Nowak, M. A., S. Bonhoeffer, and R. M. May. 1994. Spatial games and the maintenance of cooperation. *Proceedings of the National Academy of Sciences* **91**:4877–4881.

Nuismer, S. L., R. Gomulkiewicz, and M. T. Morgan. 2003. Coevolution in temporally variable environments. *American Naturalist* **162**:195–204.

O'Dowd, D. J., and E. A. Catchpole. 1983. Ants and extrafloral nectaries: no evidence for plant protection in *Helichrysum* spp. – ant interactions. *Oecologia* **59**:191–200.

Offenberg, J. 2000. Correlated evolution of the association between aphids and ants and the association between aphids and plants with extrafloral nectaries. *Oikos* **91**:146–152.

Offenberg, J. 2001. Balancing between mutualism and exploitation: the symbiotic interaction between *Lasius* ants and aphids. *Behavioral Ecology and Sociobiology* **49**:304–310.

Oliveira, P. S., and A. V. L. Freitas. 2004. Ant-plant-herbivore interactions in the neotropical cerrado savanna. *Naturwissenschaften* **91**:557–570.

Olmstead, K. L., and T. K. Wood. 1990. The effect of clutch size and ant-attendance on egg guarding by *Entylia bactriana* (Homoptera: Membracidae). *Psyche* **97**:111–119.

Osborn, F., and K. Jaffe. 1997. Cooperation vs. exploitation: interactions between Lycaenid (Lepidoptera: Lycaenidae) larvae and ants. *Journal of Research on the Lepidoptera* **34**:69–82.

Oster, G. F., and E. O. Wilson. 1978. *Caste and Ecology in the Social Insects.* Princeton: Princton University Press.

Parvinen, K., U. Dieckmann, M. Gyllenberg, and J. A. J. Metz. 2003. Evolution of dispersal in metapopulations with local density dependence and demographic stochasticity. *Journal of Evolutionary Biology* **16**:143–153.

Pemberton, R. W. 1998. The occurrence and abundance of plants with extrafloral nectaries, the basis for antiherbivore defensive mutualisms, along a latitudinal gradient in east Asia. *Journal of Biogeography* **25**:661–668.

Petersen, M. K., and J. P. Sandström. 2001. Outcome of indirect competition between two aphid species mediated by responses in their common host plant. *Functional Ecology* **15**:525–534.

Peterson, M. A. 1995. Unpredictability in the facultative association between larvae of *Euphilotes enoptes* (Lepidoptera: Lycaenidae) and ants. *Biological Journal of the Linnean Scociety*. **55**:209–223.

Pierce, N. E. 1985. Lycaenid butterflies and ants: selection for nitrogen-fixing and other protein rich food plants. *American Naturalist* **125**:888–895.

Pierce, N. E. 1987. The evolution of biogeography of associations between lycaenid butterflies and ants. In *Oxford Surveys in Evolutionary Biology*, vol. 4, ed. P. H. Harvey and L. Partridge, pp. 89–116. Oxford: Oxford University Press.

Pierce, N. E., and S. Easteal. 1986. The selective advantage of attendant ants for the larvae of a lycaenid butterfly, *Glaucopsyche lygdamus. Journal of Animal Ecology* **55**:451–462.

Pierce, N. E., and M. A. Elgar. 1985. The influence of ants on host plant selection by *Jalmenus evagoras*, a myrmecophilous lycaenid butterfly. *Behavioral Ecology and Sociobiology* **16**:209–222.

Pierce, N. E., and W. R. Young. 1986. Lycaenid butterflies and ants. Two-species stable equilibria in mutualistic, commensal, and parasitic interactions. *American Naturalist* **128**:216–227.

Pierce, N. E., M. F. Braby, A. Heath, D. J. Lohman, J. Mathew, D. B. Rand, and M. A. Travassos. 2002. The ecology and evolution of ant association in the Lycaenidae (Lepidoptera). *Annual Review of Entomology* **47**:733–771.

Pierce, N. E., R. L. Kitching, R. C. Buckley, M. F. J. Taylor, and K. F. Benbow. 1987. The costs and benefits of cooperation between the Australian lycaenid butterfly, *Jalmenus evagoras*, and its attendant ants. *Behavioral Ecology and Sociobiology* **21**:237–248.

Plantegenest, M., and P. Kindlmann. 1999. Evolutionarily stable strategies of migration in heterogeneous environments. *Evolutionary Ecology* **13**:229–244.

Poethke, H. J., and T. Hovestadt. 2002. Evolution of density-and patch-size-dependent dispersal rates. *Proceedings of the Royal Society of London Series B* **269**:637–645.

Polis, G. A., W. B. Anderson, and R. D. Holt. 1997. Toward an integration of landscape and food web ecology: the dynamics of spatially subsidized food webs. *Annual Review of Ecology and Systematics* **28**:289–316.

Ponsen, M. B. 1991. Structure of the digestive system of aphids. *Wageningen Agricultural University Papers* **91**:1–61.

Pontin, A. J. 1978. The number and distribution of subterranean aphids and their exploitation by the ant *Lasius flavus* (Fabr.). *Ecological Entomology* **3**:203–207.

Porter, S. D., and D. A. Savignano. 1990. Invasion of polygyne fire ants decimates native ants and disrupts arthropod community. *Ecology* **71**:2095–2106.

Portha, S., J.-L. Deneubourg, and C. Detrain. 2002. Self-organized asymmetries in ant foraging: a functional response to food type and colony needs. *Behavioural Ecology* **13**:776–781.

Portha, S., J. L. Deneubourg, and C. Detrain. 2004. How food type and brood influence foraging decisions of *Lasius niger* scouts. *Animal Behaviour* **68**:115–122.

Poveda, K., I. Steffan-Dewenter, S. Scheu, and T. Tscharntke. 2005. Effects of decomposers and herbivores on plant performance and aboveground plant-insect interactions. *Oikos* **108**:503–510.

Power, M. E. 1992. Top-down and bottom-up forces in food webs: do plants have primacy? *Ecology* **73**:733–746.

Prado, E., and W. F. Tjallingii. 1997. Effects of previous plant infestation on sieve element acceptance by two aphids. *Entomologia Experimentalis et Applicata* **82**:189–200.

Price, P. W. 1991. The plant vigor hypothesis and herbivore attack. *Oikos* **62**:244–251.

Price, P. W. 1997. *Insect Ecology*, 3rd edition. New York: Wiley & Sons.

Price, P. W. 2002. Resource-driven terrestrial interaction webs. *Ecological Research* **17**:241–247.

Prins, A. H., K. Vrieling, P. G. L. Klinkhamer, and T. J. De Jong. 1990. Flowering behaviour of *Senecio jacobaea*: effects of nutrient availability and size-dependent vernalization. *Oikos* **59**:248–252.

Punttila, P. 1996. Succession, forest fragmentation, and the distribution of wood ants. *Oikos* **75**:291–298.

Punttila, P., Y. Haila, T. Pajunen, and H. Tukia. 1991. Colonization of clear-cut forests by ants in the Southern Finnish Taiga – a quantitative survey. *Oikos* **61**:250–262.

Rai, B., H. I. Freedman, and J. F. Addicott. 1983. Analysis of three species models of mutualism in predator-prey and competitive systems. *Mathematical Biosciences* **65**:13–50.

Rankin, M. A., and J. C. A. Burchsted. 1992. The cost of migration in insects. *Annual Review of Entomology* **37**:533–559.

Raven, J. A. 1983. Phytophages of xylem and phloem: a comparison of animal and plant sap-feeders. *Advances in Ecological Research* **13**:136–234.

Remaudière, G., and M. Remaudière. 1997. *Catalogue of the World's Aphididae*. Paris: INRA.

Renault, C. K., L. M. Buffa, and M. A. Delfino. 2005. An aphid-ant interaction: effects on different trophic levels. *Ecological Research* **20**:71–74.

Retana, J., X. Cerda, A. Alsina, and J. Bosch. 1988. Field observations of the ant *Camponotus sylvaticus* (Hym., Formicidae): diet and activity patterns. *Acta Oecologica* **9**:101–109.

Rhodes, J. D., P. C. Croghan, and A. F. G. Dixon. 1996. Uptake, excretion and respiration of sucrose and amino acids by the pea aphid *Acyrthosiphon pisum*. *Journal of Experimental Biology* **199**:1269–1276.

Rhodes, J. D., P. C. Croghan, and A. F. G. Dixon. 1997. Dietary sucrose and oligosaccharide synthesis in relation to osmoregulation in the pea aphid, *Acyrthosiphon pisum*. *Physiological Entomology* **22**:373–379.

Ricker, W. E. 1954. Stock and recruitment. *Journal of the Fisheries Board of Canada* **11**:559–623.

Rissing, S., G. Pollock, M. Higgins, R. Hagen and D. Smith. 1989. Foraging specialization without relatedness or dominance among co-founding ant queens. *Nature* **338**: 420–422.

Ritchie, M. G., R. K. Butlin, and G. M. Hewitt. 1987. Causation, fitness effects and morphology of macropterism in *Chorthippus parallelus* (Orthoptera: Acrididae). *Ecological Entomology* **12**:209–218.

Robbins, R. K. 1991. Cost and evolution of a facultative mutualism between ants and lycaenid larvae (Lepidoptera). *Oikos* **62**:363–369.

Roche, R. K., and D. A. Wheeler. 1997. Morphological specialization of the digestive tract of *Zacryptocerus rohweri* (Hymenoptera: Formicidae). *Journal of Morphology* **234**:253–262.

Roff, D. A. 1984. The cost of being able to fly: a study of wing polymorphism in two species of crickets. *Oecologia* **63**:30–37.

Roff, D. A. 1986. The evolution of wing dimorphism in insects. *Evolution* **40**:1009–1020.

Roff, D. A. 1990. The evolution of flightlessness in insects. *Ecological Monographs* **60**:389–421.

Roff, D. A. 1994. Habitat persistence and the evolution of wing dimorphism in insects. *American Naturalist* **144**:772–798.

Rosengren, R., and P. Pamilo. 1983. The evolution of polygyny and polydomy in mound building *Formica* ants. *Acta Entomologica Fennici* **42**:65–77.

Rosengren, R., and L. Sundström. 1991. The interaction between red wood ants, Cinara aphids, and pines. A ghost of mutualism past? In *Ant-plant Interactions*, ed. C. R. Huxley and D. F. Cutler, pp. 81–91. Oxford: Oxford University Press.

Rosenzweig, M. L. 1995. *Species Diversity in Space and Time*. Cambridge: Cambridge University Press.

Ruf, C., A. Freese, and K. Fiedler. 2003. Larval sociality in three species of central-place foraging lappet moths (Lepidoptera: *Lasiocampidae*): a comparative survey. *Zoologischer Anzeiger* **242**:209–222.

Sakata, H. 1994. How an ant decides to prey on or attend aphids. *Researches on Population Ecology*. **36**:45–51.

Sakata, H. 1995. Density-dependent predation of the ant *Lasius niger* (Hymenoptera: Formicidae) on two attended aphids *Lachnus tropicalis* and *Myzocallis kuricola* (Homoptera: Aphididae). *Researches on Population Ecology* **37**:159–164.

Sakata, H. 1999. Indirect interactions between two aphid species in relation to ant-attendance. *Ecological Research* **14**:329–340.

Sakata, H., and Y. Hashimoto. 2000. Should aphids attract or repel ants? Effect of rival aphids and extrafloral nectaries on ant-aphid interactions. *Population Ecology* **42**:171–178.

Sandström, J. P., and N. A. Moran. 2001. Amino acid budgets in three aphid species using the same host plant. *Physiological Entomology* **26**:202–211.

Sandström, J., A. Telang, and N. A. Moran. 2000. Nutritional enhancement of host plants by aphids: a comparison of three aphid species on grasses. *Journal of Insect Physiology* **46**:33–40.

Scheu, S., A. Theenhaus, and T. H. Jones. 1999. Links between the detritivore and the herbivore system: effects of earthworms and Collembola on plant growth and aphid development. *Oecologia* **119**:541–551.

Schmidt, M. H., A. Lauer, T. Purtauf, C. Thies, M. Schaefer, and T. Tscharntke. 2003. Relative importance of predators and parasitoids for cereal aphid control. *Proceedings of the Royal Society of London Series B-Biological Sciences* **270**:1905–1909.

Schoener, T. W. 1986. Mechanistic approaches to community ecology: a new reductionism. *American Zoologist* **26**:81–106.

Seppa, P., L. Sundstrom, and P. Punttila. 1995. Facultative polygyny and habitat succession in boreal ants. *Biological Journal of the Linnean Society* **56**:533–551.

Seufert, P., and K. Fiedler. 1996a. The influence of ants on patterns of colonization and establishment within a set of coexisting lycaenid butterflies in a south-east Asian tropical rain forest. *Oecologia* **106**:127–136.

Seufert, P., and K. Fiedler. 1996b. Life-history diversity and local co-existence of three closely related lycaenid butterflies (Lepidoptera: Lycaenidae) in Malaysian rainforests. *Zoologischer Anzeiger* **234**:229–239.

Seufert, P., and K. Fiedler. 1999. Myrmecophily and parasitoid infestation of south-east Asian lycaenid butterfly larvae. *Ecotropica* **5**: 59–64.

Shenk, T. M., G. C. White, and K. P. Burnham. 1998. Sampling-variance effects on detecting density dependence from temporal trends in natural populations. *Ecological Monographs* **68**:445–463.

Shields, O. 1989. World numbers of butterflies. *Journal of the Lepidopterists' Society* **43**:178–183.

Shingleton, A. W., and W. A. Foster. 2000. Ant tending influences soldier production in a social aphid. *Proceedings of the Royal Society of London Series B* **267**:1863–1868.

Shingleton, A. W., and D. L. Stern. 2003. Molecular phylogenetic evidence for multiple gains or losses of ant mutualism within the aphid genus *Chaitophorus*. *Molecular Phylogenetics and Evolution* **26**:26–35.

Shingleton, A. W., D. L. Stern, and W. A. Foster. 2005. The origin of a mutualism: a morphological trait promoting the evolution of ant-aphid mutualisms. *Evolution* **59**:921–926.

Skinner, G. J., and J. B. Whittaker. 1981. An experimental investigation of interrelationships between the wood-ant (*Formica rufa*) and some tree canopy herbivores. *Journal of Animal Ecology* **50**:313–326.

Sloggett, J. J., and M. E. N. Majerus 2000. Habitat preferences and diet in the predatory Coccinellidae: an evolutionary perspective. *Biological Journal of the Linnean Society* **70**:63–88.

Sloggett, J. J., and M. E. N. Majerus. 2003. Adaptations of *Coccinella magnifica*, a myrmecophilous coccinellid to aggression by wood ants (*Formica rufa* group). II. Larval behaviour, and ladybird oviposition location. *European Journal of Entomology* **100**:337–344.

Sloggett, J. J., R. A. Wood, and M. Majerus. 1998. Adaptation of *Cocinella magnifica* Redtenbacher, a myrmecophilous coccinellid, to aggression by wood ants (*Formica rufa* group). I. Adult behavioral adaptation, its ecological context and evolution. *Journal of Insect Behaviour* **11**:889–904.

Smiley, J. T., P. R. Atsatt, and N. E. Pierce. 1988. Local distribution of the lycaenid butterfly, *Jalmenus evagoras*, in response to host ants and plants. *Oecologia* **76**:416–422.

Sober, E. and D. S. Wilson 1998. *Unto Others*. Cambridge, MA: Harvard University Press.

Solbreck, C. 1986. Wing and flight muscle polymorphism in a lygaeid bug, *Horvathiolus gibbicollis*: determinants and life-history consequences. *Ecological Entomology* **11**:435–444.

Southwood, T. R. E. 1962. Migration of terrestrial arthropods in relation to habitat. *Biological Reviews of the Cambridge Philosophical Society* **37**: 171–214.

Stachowicz, J. J. 2001. Mutualism, facilitation, and the structure of ecological communities. *BioScience* **51**:235–246.

Stacy, P. B., V. A. Johnson, and M. L. Taper. 1997. Migration within metapopulations: the impact upon local population dynamics. In *Metapopulation Biology*, ed. I. A. Hanski and M. E. Gilpin, pp. 267–291. San Diego: Academic Press.

Stadler, B. 1995. Adaptive allocation of resources and life-history trade-offs in aphids relative to plant quality. *Oecologia* **102**:246–254.

Stadler, B. 2002. Determinants of the size of aphid-parasitoid assemblages. *Journal of Applied Entomology* **126**:258–264.

Stadler, B. 2004. Wedged between bottom-up and top-down processes: aphids on tansy. *Ecological Entomology* **29**:106–116.

Stadler, B., and A. F. G. Dixon. 1998a. Costs of ant-attendance for aphids. *Journal of Animal Ecology* **67**:454–459.

Stadler, B., and A. F. G. Dixon. 1998b. Why are obligate mutualistic interactions between aphids and ants so rare? In *Aphids in Natural and Managed Ecosystems*, ed. J. M. Nieto Nafria and A. F. G. Dixon, pp. 271–278. Leon: University of Leon.

Stadler, B., and A. F. G. Dixon. 2005. Ecology and evolution of aphid-ant interactions. *Annual Review of Ecology, Evolution and Systematics* **36**:345–372.

Stadler, B., A. F. G. Dixon, and P. Kindlmann. 2002. Relative fitness of aphids: effects of plant quality and ants. *Ecology Letters* **5**:216–222.

Stadler, B., K. Fiedler, T. J. Kawecki, and W. W. Weisser. 2001. Costs and benefits for phytophagous myrmecophiles: when ants are not always available. *Oikos* **92**:467–478.

Stadler, B., P. Kindlmann, P. Šmilauer, and K. Fiedler. 2003. A comparative analysis of morphological and ecological characters of European aphids and lycaenids in relation to ant-attendance. *Oecologia* **135**:422–430.

Stadler, B., B. Michalzik, and T. Müller. 1998. Linking aphid ecology with nutrient fluxes in a coniferous forest. *Ecology* **79**:1514–1525.

Stadler, B, T. Müller, and D. Orwig. 2006a. The ecology of energy and nutrient fluxes in hemlock forest invaded by hemlock woolly adelgid. *Ecology*, **87**: 1792–1804.

Stadler, B., A. Schramm, and K. Kalbitz. 2006b. Ant-mediated effects on spruce litter decomposition, solution chemistry, and microbial activity. *Soil Biology Biochemistry* **38**:561–572.

Stanton, M. L. 2003. Interacting guilds: moving beyond the pairwise perspective on mutualisms. *American Naturalist* **162**:S10–S23.

Stanton, M. L., T. M. Palmer, and T. P. Young. 2002. Competition-colonization trade-offs in a guild of African Acacia ants. *Ecological Monographs* **72**:347–363.

Stern, D. L., and W. A. Foster. 1996. The evolution of soldiers in aphids. *Biological Review* **71**:29–79.

Stern, D. L., and W. A. Foster. 1997. The evolution of sociality in aphids: a clone's-eye view. In *The Evolution of Social Behavior in Insects and Arachnids*, ed. J. Choe and B. Crespi, pp. 150–165. Cambridge: Cambridge University Press.

Stern, D. L., S. Aoki, and D. U. Kurosu. 1995. The life-cycle and natural history of the tropical aphid *Cerataphis fransseni* (Homoptera, Aphididae, Hormaphidinae), with reference to the evolution of host alternation in aphids. *Journal of Natural History* **29**:231–242.

Stradling, D. J. 1987. Nutritional ecology of ants. In *Nutritional Ecology of Insects, Mites, Spiders, and Related Invertebrates*, ed. F. Slansky and J. G. Rodriguez, pp. 927–969. New York: Wiley-Interscience.

Strauss, S. Y. 1987. Direct and indirect effects of host-plant fertilization on an insect community. *Ecology* **68**:1670–1678.

Straw, N. A., and G. Green. 2001. Interactions between green spruce aphid *Elatobium abietinum* (Walker) and Norway and Sitka spruce under high and low nutrient conditions. *Agricultural and Forest Entomology* **3**:263–274.

Sudd, J. H. 1983. The distribution of foraging wood-ants (*Formica lugubris* Zett) in relation to the distribution of aphids. *Insectes Sociaux* **30**: 298–307.

Sudd, J. H., and M. E. Sudd. 1985. Seasonal changes in the response of wood-ants (*Formica lugubris*) to sucrose baits. *Ecological Entomology* **10**:89–97.

Szentesi, A., and M. Wink. 1991. Fate of quinolizidine alkaloids through 3 trophic levels – *Laburnum anagyroides* (Leguminosae) and associated organisms. *Journal of Chemical Ecology* **17**:1557–1573.

Takada, H., and Y. Hashimoto. 1985. Association of the root aphid parasitoids *Aclitus sappaphis* and *Paralipsis eikoae* (Hymenoptera, Aphidiidae) with the aphid-attending ants *Pheidole fervida* and *Lasius niger* (Hymenoptera, Formicidae). *Kontyu, Tokyo* **53**:150–160.

Taylor, F. 1977. Foraging behaviour of ants: experiments with two species of myrmecine ants. *Behavioral Ecology and Sociobiology* **2**:147–167.

Taylor, R. W. 1978. *Nothomyrmecia macrops*: a living-fossil ant rediscovered. *Science* **201**: 979–985.

Telang, A., J. Sandstrom, E. Dyreson, and N. A. Moran. 1999. Feeding damage by *Diuraphis noxia* results in a nutritionally enhanced phloem diet. *Entomologia Experimentalis et Applicata* **91**:403–412.

Thomas, J. A., G. W. Elmes, R. T. Clarke, K. G. Kim, M. L. Munguira, and M. E. Hochberg. 1997. Field evidence and model predictions of butterfly-mediated apparent competition between gentian plants and red ants. *Acta Oecologica-International Journal of Ecology* **18**:671–684.

Thompson, J. N. 1994. *The Coevolutionary Process*. Chicago: University of Chicago Press.

Thompson, J. N. 1997. Evaluating the dynamics of coevolution among geographically structured populations. *Ecology* **78**:1619–1623.

Thompson, J. N. 1999. Specific hypotheses on the geographic mosaic of coevolution. *American Naturalist* **153**:S1–S14.

Thompson, J. N., and B. M. Cunningham. 2002. Geographic structure and dynamics of coevolutionary selection. *Nature* **417**:735–738.

Thompson, W. R. 1924. La theory mathematique de l'action des parasites entomophages et le facteur du hassard. *Annales Faculté des Sciences de Marseille* **2**:69–89.

Tilman, D. 1994. Competition and biodiversity in spatially structured habitats. *Ecology* **75**:2–16.

Tobin, J. E. 1991. A neotropical rainforest canopy, ant community: some ecological considerations, In *Ant-plant Interactions*, ed. C. R. Huxley and D. F. Cutler, pp. 536–538. Oxford: Oxford University Press.

Tobin, J. E. 1993. Ants as primary consumers: diet and abundance in the *Formicidae*. In *Nourishment and Evolution in Insect Societies*, ed. J. A. Hunt and C. A. Nalepa, pp. 279–307. Boulder, CO: Westview Press.

Todes, D. P. 1987. Darwin's Malthusian Metaphor and Russian evolutionary thought, 1859–1917. *Isis* **78**: 537–551.

Toft, S. 1995. Value of the aphid *Rhopalosiphum padi* as food for cereal spiders. *Journal of Applied Ecology* **32**:552–560.

Tremblay, E. 1989. *Coccoidea endocytobiosis*. In *Insect Endocytobiosis: Morphology, Physiology, Genetics, Evolution*, ed. W. Schwemmler and G. Gassner, pp. 145–173. Boca Raton: CRC Press.

Trivers, R. 1971. The evolution of reciprocal altruism. *Quarterly Review of Biology* **46**:35–57.

Tscharntke, T., and R. Brandl. 2004. Plant-insect interactions in fragmented landscapes. *Annual Review of Entomology* **49**:405–430.

Tsutsui, N. D., and A. V. Suarez. 2003. The colony structure and population biology of invasive ants. *Conservation Biology* **17**:48–58.

Turchin, P. 1999. Population regulation: a synthetic view. *Oikos*, **84**: 153–159.

Turchin, P. 2001. Does population ecology have general laws? *Oikos* **94**:17–26.

Turchin, P. 2003. *Complex Population Dynamics*. Princeton: Princeton University Press.

Turchin, P., and A. A. Berryman. 2000. Detecting cycles and delayed density dependence: a comment on Hunter & Price (1998). *Ecological Entomology* **25**:119–121.

Vandermeer, J. H. and D. E. Goldberg. 2003. *Population Ecology*. Princeton: Princeton University Press.

van Ham, R., J. Kamerbeek, C. Palacios, *et al.* 2003. Reductive genome evolution in *Buchnera aphidicola*. *Proceedings of the National Academy of Sciences* **100**:581–586.

Vepsalainen, K., and R. Savolainen. 1990. The effect of interference by Formicine ants on the foraging of *Myrmica*. *Journal of Animal Ecology* **59**:643–654.

Vepsalainen, K., R. Savolainen, J. Tiainen, and J. Vilen. 2000. Successional changes of ant assemblages: from virgin and ditched bogs to forests. *Annales Zoologici Fennici* **37**:135–149.

Verhulst, P. F. 1838. Notices sur la loi que la population suit dans son croissement. *Correspondance Mathématique et Physique* **10**:113–121.

Völkl, W. 1992. Aphids or their parasitoids: who actually benefits from ant-attendance? *Journal of Animal Ecology* **61**:273–281.

Völkl, W. 1995. Behavioral and morphological adaptations of the coccinellid, *Platynaspis luteorubra* for exploiting ant-attended resources (Coleoptera: Coccinellidae). *Journal of Insect Behaviour* **8**:653–670.

Völkl, W., and K. Vohland. 1996. Wax covers in larvae of two *Scymnus* species: do they enhance coccinellid larval survival? *Oecologia* **107**:498–503.

Völkl, W., J. Woodring, M. Fischer, M. W. Lorenz, and K. H. Hoffmann. 1999. Ant-aphid mutualisms: the impact of honeydew production and honeydew sugar composition on ant preferences. *Oecologia* **118**:483–491.

Volterra, V. 1926. Fluctuation in the abundance of a species considered mathematically. *Nature* **118**:558–560.

von Dohlen, C. D., and N. A. Moran. 1995. Molecular phylogeny of the Homoptera: a paraphyletic taxon. *Journal of Molecular Evolution* **41**:211–223.

Vrieling, K., W. Smit and E. Vandermeijden. 1991. Tritrophic interactions between aphids (*Aphis jacobaeae* Schrank), ant species, *Tyria jacobaeae* L., and *Senecio jacobaea* L. lead to maintenance of genetic variation in pyrrolizidine alkaloid concentration. *Oecologia*, **86**: 177–182.

Wagner, D. 1993. Species-specific effects of tending ants on the development of lycaenid butterfly larvae. *Oecologia* **96**:276–281.

Wagner, D. L., and J. K. Liebherr. 1992. Flightlessness in Insects. *Trends in Ecology and Evolution* **7**:216–220.

Waloff, N. 1983. Absence of wing polymorphism in the arboreal, phytophagous species of some taxa of temperate Hemiptera: an hypothesis. *Ecological Entomology* **8**:229–232.

Walters, K. F. A., and A. F. G. Dixon. 1983. Migratory urge and reproductive investment in aphids: variation within clones. *Oecologia* **58**:70–75.

Ward, S. A., S. R. Leather, J. Pickup and R. Harrington. 1998. Mortality during dispersal and the cost of host-specificity in parasites: how many aphids find hosts? *Journal of Animal Ecology* **67**:763–773.

Wardle, D. A. 2002. *Communities and Ecosystems: Linking the Aboveground and Belowground Components*. Princeton: Princeton University Press.

Watt, A. D., and A. F. G. Dixon. 1981. The role of cereal growth stages and crowding in the induction of alatae in *Sitobion avenae* and its consequences for population growth. *Ecological Entomology* **6**:441–447.

Way, M. J. 1963. Mutualism between ants and honeydew-producing Homoptera. *Annual Review of Entomology* **8**:307–344.

Weisser, W. W. 2000. Metapopulation dynamics in an aphid-parasitoid system. *Entomologia Experimentalis et Applicata* **97**:83–92.

Weisser, W. W., C. Braendle, and N. Minoretti. 1999. Predator-induced morphological shift in the pea aphid. *Proceedings of the Royal Society of London Series B* **266**:1175–1181.

Wheeler, W. M. 1910. *Ants: Their structure, Development and Behaviour*. New York: Columbia University Press.

White, T. C. R. 1969. An index to measure weather-induced stress of trees associated with outbreaks of psyllids in Australia. *Ecology* **50**:905–909.

White, T. C. R. 1978. The importance of relative shortage of food in animal ecology. *Oecologia* **33**:71–86.

White, T. C. R. 1984. The abundance of invertebrate herbivores in relation to the availability of nitrogen in stressed food plants. *Oecologia* **63**:90–105.

White, T. C. R. 2004. Limitation of populations by weather-driven changes in food: a challenge to density-dependent regulation. *Oikos* **105**:664–666.

Wiens, J. A. 1977. On competition and variable environments. *American Scientist* **65**:590–597.

Wilkinson, T. L., T. Fukatsu, and H. Ishikawa. 2003. Transmission of symbiotic bacteria *Buchnera* to parthenogenetic embryos in the aphid *Acyrthosiphon pisum* (Hemiptera: Aphidoidea). *Arthropod Structure and Development* **32**:241–245.

Williams, G. C. 1966. *Adaptation and Natural Selection*. Princeton: Princeton University Press.

Wilson, D. S. 1975. Theory of group selection. *Proceedings of the National Academy of Sciences* **72**:143–146.

Wilson, D. S. 1983. The group selection controversy: history and current status. *Annual Review of Ecology and Systematics* **14**:159–187.

Wilson, D. S. 1992. Complex interactions in metacommunities, with implications for biodiversity and higher levels of selection. *Ecology* **73**:1984–2000.

Wilson, E. O. 1987. Causes of ecological success: the case of the ants. *Journal of Animal Ecology* **56**:1–9.

Wilson, E. O. 1990. *Success and Dominance in Ecosystems: The Case of Social Insects*. Nordbünte: Ecological Institute.

Wilson, E. O., and B. Hölldobler. 2005. The rise of the ants: a phylogenetic and ecological explanation. *Proceedings of the National Academy of Sciences* **102**:7411–7414.

Wimp, G. M., and T. G. Whitham. 2001. Biodiversity consequences of predation and host plant hybridization on an aphid-ant mutualism. *Ecology* **82**:440–452.

Wink, M., and P. Römer. 1986. Acquired toxicity: the advantages of specializing on alkaloid-rich lupins to *Macrosiphon albifrons* (Aphidae). *Naturwissenschaften* **73**:210–212.

Wink, M., and L. Witte. 1991. Storage of quinolizidine alkaloids in *Macrosiphum albifrons* and *Aphis genistae* (Homoptera: Aphididae). *Entomologia Generalis* **15**:237–254.

Wolin, C. L. 1985. The population dynamics of mutualistic systems. In *The Biology of Mutualism*, ed. D. H. Boucher, pp. 248–269. New York: Oxford University Press.

Wolin, C. L., and L. R. Lawlor. 1984. Models of facultative mutualism: density effects. *American Naturalist* **124**:843–862.

Wood, T. K. 1977. Role of parent females and attendant ants in the maturation of the treehopper, *Entylia bactriana* (Homoptera: Membracidae). *Sociobiology* **2**:257–272.

Wood, T. K. 1982. Ant-attended nymphal aggregations in the *Enchenopa binotata* complex (Homoptera: Membracidae). *Annals of the Entomological Society of America* **75**:649–653.

Wood, T. K. 1993. Diversity in the new-world Membracidae. *Annual Review of Entomology* **38**:409–435.

Woodring, J., R. Wiedemann, M. K. Fischer, K. H. Hoffmann, and W. Volkl. 2004. Honeydew amino acids in relation to sugars and their role in the establishment of ant-attendance hierarchy in eight species of aphids feeding on tansy (*Tanacetum vulgare*). *Physiological Entomology* **29**:311–319.

Wootton, J. T. 1994. The nature and consequences of indirect effects in ecological communities. *Annual Review of Ecology and Systematics* **25**:443–466.

Wright, D. H. 1989. A simple, stable model of mutualism incorporating handling time. *American Naturalist* **134**:664–667.

Wright, S. 1978. *Evolution and Genetics of Populations*. Chicago: University of Chicago Press.

Yamamura, N., M. Higashi, N. Behera, and J. Y. Wakano. 2004. Evolution of mutualism through spatial effects. *Journal of Theoretical Biology* **226**:421–428.

Yao, I., and S. Akimoto. 2001. Ant attendance changes the sugar composition of the honeydew of the drepanosiphid aphid *Tuberculatus quercicola*. *Oecologia* **128**:36–43.

Yao, I., and S. Akimoto. 2002. Flexibility in the composition and concentration of amino acids in honeydew of the drepanosiphid aphid *Tuberculatus quercicola*. *Ecological Entomology* **27**:745–752.

Yao, I., H. Shibao, and S. Akimato. 2000. Costs and benefits of ant-attendance to the drepanosiphid aphid *Tuberculatus quercicola*. *Oikos* **89**:3–10.

Yu, D. W. 2001. Parasites of mutualisms. *Biological Journal of the Linnean Society* **72**:529–546.

Yu, D. W., and H. B. Wilson. 2001. The competition-colonization trade-off is dead; long live the competition-colonization trade-off. *American Naturalist* **158**:49–63.

Yu, D. W., H. B. Wilson, and N. E. Pierce. 2001. An empirical model of species coexistence in a spatially structured environment. *Ecology* **82**:1761–1771.

Zera, A. J., and S. Mole. 1994. The physiological costs of flight capability in wing-dimorphic crickets. *Researches on Population Ecology* **36**:151–156.

Zera, A. J., S. Mole, and K. Rokke. 1994. Lipid, carbohydrate and nitrogen-content of long-winged and short-winged *Gryllus firmus* – implications for the physiological cost of flight capability. *Journal of Insect Physiology* **40**:1037–1044.

Zhang, Z. B. 2003. Mutualism or cooperation among competitors promotes coexistence and competitive ability. *Ecological Modelling* **164**:271–282.

Zink, A. G. 2003. Quantifying the costs and benefits of parental care in female treehoppers. *Behavioral Ecology* **14**:687–693.

Zschokke, S., C. Dolt, H. P. Rusterholz, *et al.* 2000. Short-term responses of plants and invertebrates to experimental small-scale grassland fragmentation. *Oecologia* **125**:559–572.

Species index

Plants
Artemisia ludoviciana 111
Betula pendula 104, 116
Caryocar brasiliense 82
Castanea crenata 120
Chrysothamnus viscidiflorus 154
Cirsium arvense 100
Cordia nodosa 20
Epilobium angustifolium 49, 109, 119
Fabaceae 82, 105, 113
Genista tinctoria 113
Gentiana cruciata 138, 139
Gentiana pneumonanthe 139
Laburnum gyroides 113
Lupinus albus 113
Lupinus angustifolius 113
Lupinus polyphyllus 113
Origanum sp. 138
Petteria ramentacea 113
Poa annua 114
Populus spp. 112
Populus angustifolia 166
Populus fremontii 166
Populus fremontii × *P. angustifolia* 165
Sanguisorbe officinale 139
Saraca thaipingenses 85
Senecio jacobaea 112, 114
Solidago altissima 78, 109, 155
Sophora davidii 113
Spartium junceum 113
Tanacetum vulgare 42, 85, 100
Thymus sp. 92, 139
Trifolium repens 114
Triticum aestivum 115
Vicia faba 171

Bacteria
Buchneria aphidicola 61, 89

Fungi
Marssonia betulae 116
Neozygites frensenii 156

Insects

Ants
Allomerus spp. 20
Aphaenogaster rudis 101
Attini 59
Azteca spp. 171
Camponotus spp. 78, 102
Camponotus ferrugineus 101
Crematogaster spp. 85
Dolichoderinae 57, 60
Dolichoderus spp. 102, 171
Dorymymex spp. 81
Ectatomma spp. 81
Formicinae 57, 60
Forelius foetida 81
Formica spp. 59, 102, 109
Formica altipetans 118
Formica aquilonia 160
Formica cinerea 39
Formica cunicularia 113
Formica exsectoides 138
Formica fusca 160
Formica lugubris 160
Formica obscuripes 153
Formica obscuriventris 78
Formica perpilosa 81
Formica polyctena 98, 103
Formica propinqua 165, 166
Formica rufa 98, 103
Formica rufibarbis 113
Formica subnuda 97
Formica subsericea 101
Formica yessensis 99
Iridomymex spp. 57, 72, 81
Iridomymex purpureus 84
Lasius alienus 101
Lasius flavus 102
Lasius fuliginosus 92, 125
Lasius niger 94, 99, 100, 101, 103, 113, 120, 125,
    152, 171
Lasius pallitarsis 97

212

Species index

Subject index